Loss Control
in the OSHA Era

Loss Control in the OSHA Era

CHARLES M. BINFORD

CECIL S. FLEMING

Z. A. PRUST

McGRAW-HILL BOOK COMPANY
New York   St. Louis   San Francisco   Auckland   Düsseldorf
Johannesburg   Kuala Lumpur   London   Mexico   Montreal
New Delhi   Panama   Paris   São Paulo
Singapore   Sydney   Tokyo   Toronto

Library of Congress Cataloging in Publication Data

Binford, Charles M   1927–
  Loss control in the OSHA era.

  Includes index.
  1. Industrial safety.   I. Fleming, Cecil S.,
1919–   joint author.   II. Prust, Z. A., 1924–
joint author.   III. Title.
HD7273.B56   658.38'2   74-23898
ISBN 0-07-005278-6

Copyright © 1975 by McGraw-Hill, Inc. All rights reserved.
Printed in the United States of America. No part of this
publication may be reproduced, stored in a retrieval system,
or transmitted, in any form or by any means, electronic,
mechanical, photocopying, recording, or otherwise, without
the prior written permission of the publisher.

1234567890 KPKP 78321098765

*The editors for this book were W. Hodson Mogan and Ruth L. Weine,
the designer was Naomi Auerbach, and the production
supervisor was Teresa F. Leaden. It was set in Alphatype
by University Graphics, Inc.
It was printed and bound by The Kingsport Press.*

Contents

Preface vii
Acknowledgments ix

Introduction... 1

1. The Impact of the Occupational Safety and Health Act of 1970......... 4

The Impact 4
The Key 5
The Need 5
The Main Features 6
The Point of Departure 10

2. Management Participation................................. 11

The Department Head 12
Loss Control Policy 13
Assignment of Administration 14
The Loss Control Director 15

3. The Educational Program.. 21

4. The Engineering Program.. 40

5. The Compliance Program.. 52

The Performance Evaluation Program   55
The Summary of Loss Experience   61

6. The Standard Procedure Instructions.......................... 71

    1. Emergency Reporting   73
    2. Accident Reporting   74
    3. General Housekeeping   75
    4. Safety Apparel and Appliances   76
    5. Maintenance Activities   77
    6. Chemical Laboratories   79
    7. Handling of Chemicals (General)   87
    8. Walking and Working Surfaces   93
    9. Means of Egress   99
    10. Powered Platforms, Manlifts, and Vehicle-mounted Work Platforms   102
    11. Occupational Health and Environmental Control   105
    12. Hazardous Materials   109
    13. Fire Protection   114
    14. Compressed Gas and Compressed Air Equipment   117
    15. Material Handling and Storage   118
    16. Equipment Guarding for Points of Operation and Machine Drives   122
    17. Hand and Portable Power Tools   126
    18. Electrical Wiring, Apparatus, and Equipment   128
    19. Operation of Motor Vehicles   130
    20. Handling and Storage of Explosives   133
    21. Claims Administration   138

Where Does It All End?..145

Glossary of Terms   147
Index   157

Preface

God may have created the universe in six days, but to establish a workable loss control program within all the procedural and economic parameters of your company may take a little longer. The decision to establish a formal program sounds easy, and with the emphasis these days on safety through federal and state legislation, easy or not, the decision must be made. Once made, whether your company is very small, very large, or in-between, the decision must be based on your management willingness to apply all the proven principles of good management to the program. It must be planned, organized, motivated, and controlled. If these principles sound familiar it's because they are the same elements of management that produce profit and create the atmosphere in which the overall company objectives are attained.

So . . . . in the beginning the loss control program is created. Having been so created it is implemented. Implemented efficiently, it fits the company objectives and its success, immediate and long term, contributes to the profit—might even be the prime reason for it. Even more important, it creates an atmosphere in which your company is fulfilling all those social responsibilities so hard otherwise to attain when profit, no longer being a dirty word, is the prime objective. Why does the successful loss control function do all these things? People are happier, people are healthier, people are wealthier, people—people—people. That's what it's all about, isn't it? Even in these days of machines doing all the things people used to do, there is no business that has eliminated the people; and while an oilcan may make a machine happy it doesn't do much for the people. Okay, the philosophy is established, the management is in front of it (not behind it), the risk manager needs it—let's get on with it!

Charles M. Binford
Cecil S. Fleming
Z. A. Prust

Acknowledgments

The authors gratefully acknowledge these individuals and organizations for their contributions to this manual:

The American Society of Safety Engineers for use of the format of the article "Qualitative and Quantitative Analysis" as authored by Messrs. Roman F. Diekemper, Marsh McLennan Co., and Donald A. Sparks, Pettibone Corp.

The Hartford Insurance Group for previous technical advice as furnished by Mr. John Pickens, Vice President, Engineering Department.

The Tucson Realty and Insurance Company for financial and moral support as furnished by Mr. Albert Gibson, President.

The terminology used in our manual is common to the safety profession. Our glossary of terms, for uniformity, is intended to coincide with the *Dictionary of Terms Used in the Safety Profession,* published by the American Society of Safety Engineers, December 1971.

Loss Control
in the OSHA Era

Introduction

The primary purpose for developing *Loss Control in the OSHA Era* is to offer some meaningful reinforcement to the people responsible for loss control. Also, it is our intent to assist them in initiating some positive plans for the control of conditions that could produce losses and adversely influence the objectives of their respective companies.

Our experience with educational institutions and commercial and industrial enterprises discloses a wide variation in management's concept of loss control. Executive management assumes, in some instances, that their loss control requirements are fulfilled if they buy the amount of insurance suggested by an insurance agent and if they simply assign loss control as an additional duty to an employee in the personnel, physical plant, or security department. There are other instances where an entire department is responsible for the loss control function, including insurance management, supervision of medical facilities, inspections, performance evaluation, accident investigation, and basic and interim training for employees.

Loss Control in the OSHA Era is intended as a manual to afford a ready reference and point of departure for the plant manager and administrator for loss control in all types of industrial, manufacturing, construction, and business enterprises.

The term "loss control" is identified here as an all-inclusive function, considering the control of conditions that can contribute to or be totally responsible for the interference with orderly progress to attain the objectives of the company. Losses are generally recognized as the result of an accident or an unintended interruption of an activity, involving damage to equipment, facilities, materiel and, infrequently, injury to people.

The full impact of a loss is not always recognized by management. However, when there is serious property damage or injury to personnel, it is not uncommon to record certain related data to recover a portion of the loss through insurance. Also, some changes in operations may be instituted to correct the apparent cause of the accident that resulted in loss. There are losses of varying degree and frequency, in all business activities every day, that are not identified as controllable. They are, more often than not, accepted as the normal course of operations. The *recognition* of the conditions that lead to personal injury, damage to equipment and facilities, and loss of use of a facility is not a difficult task. The installation of appropriate *control* measures to reduce or eliminate the probability of loss does, however, present a challenge. All persons of supervisory status must unite in an effort to implement the controls with uniformity.

Every loss, regardless of extent, indicates a breakdown in management control at the supervisory level most closely associated with the activity where the loss occurred. Losses resulting from similar conditions may not occur with sufficient frequency to alert supervisors to the need for reinforcing their control. Also, the severity of loss may vary from light to moderate without triggering an alarm to indicate something is wrong. However, a "once in a hundred years" type of occurrence resulting in severe loss may trigger an all-out effort to identify the conditions responsible and to install corrective measures. The corrective measures are usually concerned only with this occurrence and other facilities with similar exposures. However, it is common for the controls to "die on the vine" without someone responsible to keep them alive.

The total concept of loss control should involve *more* than the brush-fire fighting tactics often associated with the individual loss occurrence and certainly should be *less* than all-out controls for conditions that do not exist or may never be encountered.

The treatment of exposure where there is probability of loss should be the responsibility of a supervisory person. However, there must be written guidelines which provide a reasonable reference and sufficient "know-how" by the loss control department to develop the supervisor's skill for implementing the guidelines.

We believe that the fundamentals set forth in *Loss Control in the OSHA Era* will emphasize the areas that can be structured for successful loss control in almost any activity. The manual is couched in simple everyday language. We have focused on the recognized normal progression for developing a workable program to minimize the probability of loss while protecting the assets of the company.

The standards from the *Federal Register,* 29CFR1900, July 1973, were used as the basis for establishing primary guidelines for persons with supervisory status. Although a supervisor does not have primary concern for construction or manufacturing specifications, the Standard Procedure Instructions are meaningful and can be implemented during the normal course of operations as part of the management function. The published standards for construction and manufacturing of equipment should be the concern of the loss control engineer, purchasing agent, and executive management. The methods and procedures related to the workshop conferences for developing supervisory skills, measurement standards for loss control performance evaluation, and loss summary analyses are recognized as reliable old standbys. They have been in use with some variation by the loss control professional over the years.

CHAPTER ONE

The Impact of the Occupational Safety and Health Act of 1970

THE IMPACT

On December 29, 1970, the President signed into law the Williams-Steiger Occupational Safety and Health Act (OSHA) of 1970. This law became effective on April 28, 1971. It provides that each employer has the basic duty to furnish his employees both a type of employment and a place of employment which are free from recognized hazards that are causing or are likely to cause death or serious physical harm. The new law with its safety and health standards, considered by the Secretary of Labor to be more effective than old safety laws, encompassed the previous laws with respect to safety and health standards. These included the Walsh-Healy Act of June 30, 1936, the Service Contract Act of 1965, the Federal Construction Safety Act of 1969, and the National Foundation on Arts and Humanities Act of 1958. The intent of OSHA is to assure safe and healthful working conditions, within definite guidelines, for working men and women.

The congressional findings disclosed a substantial financial burden upon interstate commerce arising from personal injuries and illnesses. Therefore, the congressional power was exercised in the interest of preserving our human resources and reducing this financial burden. Initially, the effort appears to have been directed toward working conditions provided by the employer and working practices of the employee. This is apparent by the published standards and the compliance inspections at the various workplaces. Training in recognition, avoidance, and prevention of hazards (RAP) is offered to the contractor-employer for employees of the construction industry. A subsequent training program is contemplated for employees in other industries.

THE KEY

The key sections to the Occupational Safety and Health Act are Section 5, "Duties," and Section 6, "Occupational Safety and Health Standards." This is apparent by the penalties designated in Section 17 for an employer's noncompliance with these sections.

The requirements for compliance by the employers and the employees are also set forth in the official poster that was provided to every employer with an Internal Revenue Service number. The poster, additionally, identifies the rights of employees to request inspections by the Department of Labor and some of the penalties in fines and/or imprisonment that can be imposed upon the employer for violation or noncompliance with the standards. The absence of any notice of penalty to the employee for noncompliance with safety and health standards, rules, regulations, and orders issued under the act and applicable to his conduct is very conspicuous. However, the employer, having validated that he or she in fact did provide the necessary safety and health training to control work practices, may have substantial ground for removal of an employee from the payroll for noncompliance with employee duties outlined in Section 5 of the act.

THE NEED

Prior to the act, the voluntary programs maintained by employers for safe and healthful working conditions varied from excellent to virtually none. The nature and scope of the activity had some influence on the physical arrangement of facilities and equipment. Also, the materials involved in the activity determined some work practices necessary for continuity of operations. The loss experience of an activity or group of activities with similar operations influenced some employers to develop control programs with specific rules and procedures. These programs were determined by the most recent type of loss or a recent survey by a representative of the company having a vested interest in the operation.

The gradual recognition of the influence of continued losses on profit yield has caused some multilocation corporations to develop a risk management department. Here, a continuity of effort is maintained to achieve the corporate objectives with fewer losses. The availability of the insurance consultant and technical services also contributes to the success of this effort.

A large portion of the industries, including construction, remain without the semblance of a workable program for safe and healthful working conditions for working men and women. Many employers still do not have sufficient knowledge of the Occupational Safety and Health Act.

THE MAIN FEATURES

The number of citations for violations of standards, absence of official posting of occupational injury and illness records, and the summary of experience from the effective date to the present time may be indicative of ignorance rather than lack of concern for the law. We have outlined the three important features of the Occupational Safety and Health Act, for a better understanding of an estimated 10 percent of the problems confronting employers in their attempt to control losses.

Requirements for Employers

1. Furnish a place of employment free of hazards which cause illness, injury, or death to employees (so-called "general duty" clause).

2. Comply with all standards adopted by the Secretary of Labor. These include: interim standards effective August 27, 1971, although some interim standards are effective February 1972; permanent and

emergency temporary standards. The complete OSHA Regulations 29 CFR 1900 including standards, current to July 1973, are published in the *Federal Register.*

3. Permit government inspectors to enter workplaces without delay during regular working hours or at other reasonable times.

4. Conduct periodic inspections for safety and health hazards, if required by the standards.

5. Post notices to keep employees informed of their rights and duties including provisions of applicable standards.

6. Maintain records of all work-related injuries, illnesses, death, and exposure of employees to toxic substances and harmful physical agents.

7. Provide physical examinations for employees to determine the extent of exposure to toxic substances or potentially harmful physical agents which have exceeded permissible limits. (No standards or regulations have been issued for this phase as yet.)

8. Post copies of citations of violations at or near each place a violation occurred.

9. Provide employees with protective equipment where required by applicable standards.

10. Employers will be exposed to both mandatory and/or discretionary penalties for violations of any of the above.

Employee Responsibilities

There is only one requirement that the act places on employees and that is the duty to comply with *OSHA standards and rules which are applicable to his own conduct.* The act, instead of creating employee requirements, does in fact create employee rights.

Under the law, employees and/or their authorized representatives have the right to

1. Have a place of employment free from safety and health hazards.

2. Request an inspection from the Department of Labor if they suspect a violation exists.

3. Accompany the federal inspector (or state inspector if a state plan is approved) on his inspection rounds.

4. Receive notification from the inspector, in writing, whether or not a violation exists.

5. Be apprised of all existing hazards (labeling).

6. File written objections to and request a hearing on a proposed standard.

7. Report suspected violations to the government without employer retaliation, and with legal recourse if retaliation occurs.

8. Protest length of time given by the government for the employer to correct a violation.

9. Bring legal action against the Secretary of Labor for failing to take action on a violation that results in illness, injury, or death.

10. Supply the Secretary of Labor with information that may be used to develop a new standard.

11. Observe monitoring of toxic substances and harmful physical agents, and have access to records of such monitoring.

12. Be represented on the National Advisory Committee on Occupational Safety and Health, the National Commission on State Workmen's Compensation Law, and Standards Advisory Committee that the Secretary of Labor may appoint (applies mostly to union organization).

The Compliance Officer's Inspection

When an inspection takes place, under the law it should proceed as follows:

1. An inspection is initiated by an employee and/or his representative or simply on the initiative of the labor department.

2. Advance notice of inspection without the labor department's authorization is not only prohibited but penalized with a fine or imprisonment of not more than $1,000 or six months, or by both.

3. Credentials must be presented to the owner, operator, or agent in charge.

4. The inspector is entitled to enter without unreasonable delay.

5. An inspector may gain entrance by presenting credentials to any employee if minimum delay fails to produce the agent in charge.

6. The investigation and inspection must be conducted during regular working hours or at other reasonable times.

7. The employer and an authorized employee representative may accompany the inspector during the inspection.

8. In the absence of an authorized employee representative, the inspector may consult with a reasonable number of employees.

9. The inspector will check the place of employment and all perti-

nent conditions, structures, machines, apparatus, devices, equipment, and materials.

10. The inspection may include a possible hazard reported before or during an inspection.

11. A citation will be issued for any violation.

12. The citation will detail the violation and the standard violated and set an (alleged) reasonable time for correction.

13. The citation must be posted at or near the place where the violation took place.

14. An employee and/or his representative may appeal the time allowed to correct the violation.

15. Within a reasonable time after a citation is issued, the employer must be notified of the proposed penalty to be assessed.

16. The employer may appeal the citation or proposed penalty for violation (within 15 working days).

17. Inspectors may not necessarily be entitled to enter areas of a plant where operations or techniques involving trade secrets take place.

We have indicated to you that OSHA was involved with an estimated 10 percent of the employer's loss control problem. However, since December 30, 1970, a shift in emphasis has been reflected. We now believe that the majority of the loss control effort of those companies having some form of risk management program is devoted to compliance with the Occupational Safety and Health Act of 1970. During this shift in emphasis, companies seemed to accept the "easy way out" by purchasing insurance for certain risks. They were not sufficiently concerned with the probability of loss or with the existing or needed controls to assume some of the risk.

The loss control effort should encompass all areas of activity with specific attention given to each of these risk categories:

1. Workmen's compensation (employee's occupational injury and disease).
2. Public liability (property damage and bodily injury).
3. Automobile liability (vehicles common to highway travel).
4. Automobile physical damage and cargo losses.
5. Employee travel accidents (transportation).
6. Burglary and robbery.
7. Fidelity and surety bonds.
8. Fire and allied exposures.

9. Business interruption.
10. Valuable papers.
11. Accounts receivable.
12. Contractor's equipment (common to off-highway use).
13. Special equipment (inland marine).
14. Group life, accident, and sickness plans for employees.
15. Completed operations.
16. Products liability and consumer safety.

The risk management personnel are in the best position to review operations and determine the effectiveness of management controls. Studies of various activities with attention directed to facilities, processes, and people should reflect the degree of supervision needed for optimum equipment performance and employee productivity.

Risk improvement studies take into consideration all aspects of the operation. There will be many times where corrective measures outlined in a Standard Procedure Instruction will effect improvement in more than one risk area. For example, specific guidelines for employee operating practices will almost without exception prolong the useful life of equipment and reduce interim maintenance requirements. Risk improvement should not be limited or geared specifically to the law, but should encompass the entire operation to attain company objectives within the law.

POINT OF DEPARTURE

Many of the employers not in compliance with the Occupational Safety and Health Act of 1970 would not hesitate to restructure their operations if some assistance were available to initiate some independent action. The basics offered here generally apply. However, the risk manager must tear down, restructure, and refine the basics into a program that is identified with his company. The risk manager should take advantage of all assistance and technical guidance available to him from sources other than the occupational safety and health administration. He can restructure his operations so they comply with the requirements of the law and, at the same time, enhance his program. He can expect improved employee performance, more reliable operation of tools and equipment, less maintenance on a demand breakdown basis, and an improved profit picture.

CHAPTER TWO

Management Participation

The extent of management participation for controlling losses is measured by how well management is kept informed of what is happening within the activity. Executive management will, almost without exception, learn of a major breakdown, serious fire involving a major structure, fatality, serious injury, armed robbery, or major vehicle crash involving a facility owned vehicle. Unfortunately, it is not uncommon for management to be only generally aware of the loss picture in the overall activity.

Contrary to the National Safety Council's emphasis declaring the supervisor as the key man, we urge you to consider the topmost executive as the key man for your loss control effort. We agree with the concept of the supervisor being the key within the scope of his area of operations and responsibility, and hopefully he does, within those limits, administer policy and procedures as furnished to him. However, he does not usually view the "big picture" and is not in a position to formulate policy or procedures to generally control conditions that produce losses.

The occasional presence of the topmost executive of the activity at the institutional or corporate safety committee meeting is not harmful, and may produce additional viewpoints not previously considered. Management should be vitally interested in what the safety department and the committee members have identified as problems confronting the activity. The solution to the problem and application of corrective measures to arrive at the solution should be of special interest to executive management. The interest is more than short term when you recognize that the principles of management for planning, organizing, and controlling conditions that may influence the objectives of the activity include the loss control function.

Management is receptive to proposals that will improve the facilities, processes, and development of people. The proposal that will do one or more of these and influence a budgetary saving is not only welcome, but is enthusiastically sought after by those who can "make it happen."

THE DEPARTMENT HEAD

The department head in reality is the administrator of all policy and procedures which concern all operations within his or her activity area. The interpretation and implementation of any guidelines and regulatory controls affecting any portion of the department operations may be modified to control conditions for any foreseeable situation within the department. The extent of the modification, however, should not detract from or weaken the original control standard.

Enthusiastic participation by the department head depends upon the effective presentation of the control program by executive management. A random sampling of various educational institutions and industrial activities discloses that a blanket presentation of a control program to department heads does not produce a uniform all-out effort to perform within the written guidelines. There will be total enthusiastic acceptance of the control program by some, and quiet, partial, and sometimes total rejection by others. The need for the guidelines and controls is not always identified by the department head, and on these occasions the control program is transmitted to the general supervisor with the comment: "Do what is necessary to comply." Here is a direct blow to the management effort because the supervisor may not have the total experience to recognize the conditions to be controlled in each of the operations performed within the department.

It is important for the department head to know the degree of exposure to loss for every risk category in every department. Then he can effectively utilize the services of the risk management people and his supervisors to implement the control program in order of importance.

The corporate executive cannot afford the luxury of formulating and furnishing a control program to management and then expect that an uninterrupted continuity of transmission will produce the desired result. He cannot afford to consider that implementation of the control program is wholly the responsibility of the department head. There must be unity of effort at the supervisory level to reflect both the satisfactory arrangement of equipment and use of facilities as well as the efficient performance of people. The lack of acceptance or departure from the control program can be readily identified by the knowledgeable department head and the loss control director by the presence of poor work practices and substandard job conditions.

LOSS CONTROL POLICY

Involvement of executive management in the loss control program begins with the *corporate policy for loss control.* This policy is the keystone of the loss control foundation. Aside from being furnished to all persons with supervisory status, it is emphasized in orientation and training sessions. The example presented here gives the company an opportunity to state the acceptance of responsibility for loss control. Also, it specifically shows how compliance with the loss control measures is in keeping with the company policy.

MANAGEMENT CONTROL FOR LOSS CONTROL—
A POLICY OF COLORSCOPE, INCORPORATED

The loss control policy for all our corporate activity shall include the necessary guidelines for all staff for a total loss control effort. The guidelines are intended to control conditions that can be responsible for loss.

It is our intent to furnish each individual involved in any way with company objectives with facilities that are free from recognized hazards which are or can be responsible for injury, illness, and loss of properties. In this connection, each individual is expected to comply with all identified life, safety, and health standards and all rules, regulations, and orders pursuant to our policy for the prevention of losses of all types.

The prevention of injury and illness to our people shall certainly be uppermost in our endeavors. However, the concept of loss control through supervisory control shall extend beyond the normally accepted program for illness and injury. There can be occurrences related to property, materiel, and people that can have an adverse influence on the objectives of the company. The purpose of our loss control effort, therefore, is to identify these conditions and install the appropriate corrective measures prior to the occurrence.

The responsibility for loss control cannot be delegated as a staff function, but must be accepted by each member of the staff. The extent of education, engineering, and compliance for loss control in all areas of activity will be reflected in measurable experience records and the enthusiastic support of all our people.

Dated: May 10, 1973   Jeremiah Pinckney
President

ASSIGNMENT OF ADMINISTRATION

The assignment of responsibility for the administration of the loss control policy with certain specific objectives extends the management effort into all areas of activity. The position of loss control administrator may evolve as an additional duty in the smaller business enterprise. However, identifying both the administrator and the related responsibility affords a directional stability to the supervisory force. The administrator should be a problem solver with a good knowledge of a firm's facilities, processes, and people. He should also have entree to all departments with little more than ordinary protocol. He should be considered as the source of assistance in improving loss control performance. The example that follows provides the emphasis for continuity of management effort.

ADMINISTRATION FOR LOSS CONTROL

Mr. Armer Biltwell, loss control director, shall be responsible for directing the efforts of the entire staff through the normal supervisory channels to properly analyze and control losses and loss exposure to facilities, equipment, and people. All staff having supervisory status shall be responsible for implementing our loss control guidelines as they apply to each supervisor's respective activity.

15 / Management Participation

OBJECTIVES

The specific objectives of loss control for Colorscope, Incorporated are:
 1. To demonstrate that effective loss control is an integral part of supervisory control.
 2. To establish that the performance of people provided with instructional guidelines shows fewer losses.
 3. To emphasize that each person with supervisory status is responsible for controlling the conditions which may affect people and property in the supervisor's specific area of activity.
 4. To afford a uniform method for reporting, recording, and investigating losses, and to identify and cause to be installed the appropriate corrective measures in all areas with similar exposure to loss.
 5. To present an organized plan to improve the performance of people and the arrangement of facility before a loss is experienced.

These objectives, when accomplished and coupled with the guidelines to be published in the company loss control manual, will provide the base on which our loss control program is built. The loss control program, if effective, will be a major factor contributing to the overall company objectives: attainment of reasonable profit and fulfillment of social responsibility.

Dated: May 10, 1973

Jeremiah Pinckney
President

THE LOSS CONTROL DIRECTOR

The title "loss control director" is considered to be appropriate in connection with the safety management function, since all resources at a particular facility are available to aid him in resolving almost any condition that could contribute to a loss.

The loss control director should not be the activity's police officer, nor should he circumvent the supervisory responsibility for job conditions and work practices. However, any study of an activity, department, or function should include observation of job conditions and work practices.

The loss control director should not be required to investigate all losses. He should assist in the investigation of conditions that produced the loss and coordinate the efforts of department heads to install the appropriate corrective measures wherever similar conditions exist and could produce additional losses. Occurrences involving serious loss, for which no identifiable controls have been established but which warrant special investigations to specifically identify the problems, may be encountered by the supervisor or department head.

The loss control director must have the ability to communicate with others. Also, he should be outgoing and have a sincere desire to accomplish his goals with the coordinated efforts of others.

The loss control director should be a problem solver. He devises the ways and means to assist the department head in getting the job done safely. He does not set up roadblocks while paving the way for loss control.

It is not our intent to specify the position of the loss control director in the organizational structure. We do, however, emphasize the importance of having executive management identify the relationship of the loss control director's position to all levels of supervisory management for optimum unity of effort. In this connection, the loss control director should have entree into each department with a minimum of protocol. A prior personal contact, telephone call, or minimal written communication should be sufficient to arrange for a visit and otherwise enable the department head to request a scheduled loss control study for any portion of the department or facility where the need for assistance is identified.

We emphasize that any activity area having an unusual loss frequency for what may be deemed ordinary exposures should not be the sole concern of the supervisor. We equally emphasize that the loss control director should not assume the "take charge" attitude. The supervisor must remain the responsible person. A supervisor should welcome any help that will improve his performance. He must build upon or reinforce the controls he has already established, and it is most important that the loss control director work with him.

The loss control director who limits his concern to identifying substandard or less than desirable conditions without offering to management a definite plan of action with reasonable alternatives is identified as "Management Headache No. 13." The accompanying symptoms are disgruntled supervisors, employee unrest, absenteeism, schedule delays, and buildup of immunity to interference with long-standing practices, right or wrong, on too many occasions. Unfortunately, there is no effective analgesic remedy for this malaise, and a frequency of such headaches may result in drastic action by management to remove the source. That is, management will eventually lose patience with such a performer and sever his connection with the firm.

It is necessary for management to maintain a definite policy to control losses of all kinds. Moreover, this policy should be updated annually. A current policy is a basic tool for the loss control department.

The loss control director who is involved in the preliminary stages of development for any additional activity should evaluate the project planning for each risk category. All regulatory controls, standard procedures, and methods should be reviewed to build in the necessary features for optimum performance. The presentation of "too little too late" to management yields loss of face, funds, and facility and could contribute to future operating losses. We have prepared a position description for a loss control director and offer it here for your consideration. We emphasize the managerial qualities of planning, organizing, and controlling.

POSITION DESCRIPTION: Loss Control Director

ADMINISTRATIVE
Group   Code

FUNCTION AND GENERAL SCOPE

The loss control director reports to the vice president for all business affairs and is responsible for administering the policy of the corporation for loss control. This and certain other supervisory positions constitute a part of the corporate risk management team.

All administrative and department heads from whom reports are required in connection with the administration of the loss control program and all members of the staff of the loss control director (e.g., clerks, specialists, and technicians) must report to the director. The amount of direct assistance needed by the loss control director to discharge his responsibilities properly depends upon the degree of responsibility, the size and operating policies of the facility, and the type of operations. The director collaborates with the department heads and safety committee in planning loss control objectives consistent with the policies and guidelines established by the corporation.

The incumbent spends a major part of his time in training, guiding, and counseling both in the office and in the field, for the purpose of improving loss control performance within and among the several departments. He is responsible for conducting studies as may be requested and needed by the various departments.

The director is responsible for expense forecasts and observance of an effective expense control program encompassing facilities, processes, and personnel in consonance with good loss control management. He or she is also responsible for providing the executive management of the enterprise with a summary of the total loss experience on a quarterly basis and for reviewing such losses with the department heads and safety committeee for the purpose of giving direction to the loss control program.

The loss control director is accountable for the following:

1. Providing sufficient information of a staff and technical nature to enable an evaluation of loss potential in all departments.

2. Providing necessary proposals to department heads to reduce the probability of loss that has been identified during studies and surveys.

3. Providing the proper leadership, guidelines, and counseling so that supervisory persons may be properly trained and motivated for improved performance.

4. Actively participating on the risk management team. Establishing and maintaining liaison with:
   a. Insurance carriers.
   b. Loss control consultants (membership in safety engineering and industrial hygiene societies).
   c. The state industrial commission, fire marshal, and director of occupational safety and health.
   d. The insurance services office of the state.
   e. Local police and fire departments.
   f. National Safety Council and the state Safety Council.

5. Advising executive management of new and unusual situations that may be encountered which may affect the realization of the objectives.

TYPICAL DUTIES

1. Maintains all related records required to support the program.

2. Acts as a recorder for the central safety committee.

3. Submits periodic reports on the loss control status of each facility.

4. Maintains the accident record system, makes necessary reports, personally investigates fatal or serious accidents, secures supervisor's accident reports, and checks corrective action taken by the supervisor to eliminate accident causes.

5. Formulates emergency plans.

6. Keeps each director of physical resources informed of maintenance conditions that are likely to produce losses.

7. Collects statistical data, studies accident causes, and recommends corrective measures which coincide with written standards.

8. Formulates, administers, and makes necessary approved changes in the loss control program; keeps abreast of OSHA requirements.

9. Acts in an advisory capacity on all matters pertaining to safety as required for the guidance of management.

10. Coordinates the activities of others so that the loss control program

will be efficiently operated; he may delegate certain specific duties to others on his staff.

11. Reviews designs of new facilities to be owned and occupied by the corporate activity and makes suggestions for their safe operation; he should be consulted on safety features for any new building.

12. Periodically reviews the efficiency of the entire conservation program; reviews details of the plan with supervisors.

13. Reviews security measures for life safety which are concerned with vehicular and pedestrian traffic and makes recommendations for corrective action.

14. Initiates semiannual inspections of portable fire-fighting equipment, fire alarms, fire evacuation plans, evacuation routes and exits to be used, and submits appropriate reports.

15. Reviews standards for safety equipment, tools, and working conditions; provides check lists to supervisors for inspection of safety belts, ladders, ropes, rubber goods, and tools; observes crews at work and identifies performance with respect to standards, such as failure to use goggles, rubber gloves, and other safety devices; evaluates the application of controls for conditions concerning ditches and tunnels, and recommends reinforcement, where necessary, for use of braces to prevent cave-ins. *He does not circumvent supervisory responsibility.*

16. Periodically visits plant facilities to inspect equipment and installations; studies work areas to determine if safety regulations are being observed and if accident prevention devices are being used; makes sure that working areas are tested for toxic fumes, explosive gas-air mixtures, combustible materials, and other hazards.

17. Inspects machinery to determine places where danger of injury exists. Recommends installation of guards on machine drives and points of operation to comply with regulatory controls.

18. Randomly inspects premises for fire hazards and adequacy of fire protection and reviews emergency plans with persons of supervisory status.

19. Has good knowledge of federal, state, and local laws, ordinances, or orders bearing on occupational safety and health, and measures the extent of compliance.

20. Reviews standards for safety equipment to be used by plant personnel.

21. Inititates investigations to be made of all accidents involving damage to corporate property at all locations and in all areas of operation; makes reports to the risk management claims-handling department.

22. Procures and distributes safety promotional items such as booklets, slogans, pamphlets, handouts, and posters.

23. Conducts safety and loss control classes.

24. Assists supervisors in their safety inspections, accident investigations, and safety training obligations.

25. Initiates activities that will stimulate and maintain employee interest.

26. Attends high-level safety and loss control seminars and meetings, and keeps the general manager and staff informed on all matters related to loss control and safety.

27. Maintains outside professional contacts to exchange information and keep program up-to-date.

28. Maintains liaison with all departments and allied organizations.
   a. With director of physical resources for fire protection, installations, protection, maintenance of fire protection equipment, and housekeeping.
   b. With finance or accounting on loss control surveys of monies and securities.
   c. With the public fire department and loss control consultants for facility layout, location and type of hazardous processes, coordination of forces, location and type of fire-fighting equipment, fire doors, fire walls, defense plans, and engineering plans.
   d. With the corporate security division for protection of persons and of property.

DESIRED QUALIFICATIONS

1. Superior managerial ability and personnel supervisory skills.
2. Comprehensive knowledge of physical security, traffic planning, and safety planning.
3. Working knowledge of modern office procedures.
4. Should be a university graduate with five years of experience in the general safety field with firsthand knowledge of and experience in some related field, and should have some specialized training in industrial safety.
5. Resourcefulness, with ability to work with people from the operating level to the topmost executive office.

CHAPTER THREE

The Educational Program

The historical facts concerning loss experience for educational, industrial, and commercial enterprises indicate that a large percentage of losses result from a failure of an employee to follow established safe practices. This is often referred to as "an unsafe act" on the part of the individual. Hardly ever is the supervisor charged with contributing to the accident by permitting the employee to perform in the identical "unsafe" manner many times before a serious accident occurred. In essence, the supervisor contributed to the loss if the "unsafe act" was a matter of past practice rather than an error in judgment on the part of the individual at the time of the occurrence.

A conscientious program founded on simple fundamentals and procedures will succeed in providing the necessary loss control education for all faculty, staff, and casual employees. It will encompass: (1) recruitment of personnel, (2) preemployment physical examination, (3) orientation and instruction, (4) assignment and demonstration, (5) performance review, (6) progressive educational instruction, and (7) development of supervisory skills.

1. Recruitment of Personnel

The selection and placement of staff shall be accomplished in a manner that will ultimately develop for each position an individual of highest caliber who will contribute to the realization of company objectives.

Each prospective employee is expected to be a participating member of the company and shall be examined for ability, conduct, and verifiable past performance, as a base for progressive development.

2. Preemployment Physical Examination

All prospective employees, after being selected, shall be subjected to a physical examination by a competent physician. In addition to the standard physical, because of the exposure in some areas of the activity, the examining physician may have to pay special attention to certain physical requirements. These requirements may involve dexterity, climbing, lifting, extended walking, stooping, etc.

3. Orientation and Instruction

The new employee shall be properly introduced to the company's objectives, functions, and projected goals. He or she has demonstrated qualities that are attractive to us; now we shall demonstrate that our company is worthy of continued alliance. Preliminary instruction relating to general regulatory controls shall be conveyed to the individual. The instruction shall be specific and in accordance with accepted practices normally expected of all persons involved in our activity.

4. Assignment and Demonstration

The supervisor shall accompany the employee to the facility or unit where he or she is expected to work and is assigned responsibilities. The responsible supervisor shall acquaint the employee with the facilities, equipment, and materials that he or she will normally encounter, and arrange for controlled performance. A frequent review of employee progress and ability is necessary until assurance of reliable performance is recognized.

5. Performance Review

The person with supervisory status and infrequently personnel of the loss control department shall conduct occasional studies of all employees' performance, taking note of ability, behavior, and effectiveness. A program of improvement shall be projected whereby corrective measures can be collective as well as individual. Improvement programs shall include instructional training meetings, review of Standard Procedure Instructions, modifications, and new standards as they are established. Corrective measures, as required of the individual staff member, shall be conveyed in a manner to improve employee performance and not detract from favorable attitude or other satisfactory attributes.

6. Progressive Educational Instruction

The constantly changing technological society has some influence on most industrial and commercial enterprises. The ability of the company to adapt to these changes successfully depends upon how well staff and supervisory persons have kept abreast of operational requirements. To this end a degree of flexibility shall be maintained within the framework of the organization of personnel engaged in any activity connected with the company. Additional training of individuals who may be prospects for more responsible positions within the operation shall be made available. The training shall be progressive and utilize all available media to enhance the individual's value to the company. General educational instruction available through the library, bulletins, technical papers, posters, and special courses and seminars shall be incorporated into this program of education for loss control.

7. Developing Supervisory Skills

The ultimate objective of the supervisory development program for loss control is to reduce claim costs by preventing losses. This enables you to purchase adequate insurance at lowest possible cost. It will also favorably influence budgetary requirements because the cost of insurance is based, to a very large extent, upon the amount of money that the insurance companies pay out in claims on behalf of the insured.

PROGRAM FOR DEVELOPING SUPERVISORY SKILLS

Our experience points out that most losses are the result of operating conditions which the supervisor of the department often is in the best position to control. If the supervisor is trained to look at accidents in the same way he considers all other aspects of operations in his department, he will be in a better position not only to control accidents, but also to improve his total performance as a supervisor.

The development of the supervisory skills needed to deal with accident prevention is unique and, at the same time, very effective. This is why making available the tools by which the supervisor can better control accidents will have a favorable influence on the total operations. One way in which these skills are provided is in a series of workshop sessions designed to help the supervisor achieve better accident control through operations control.

The specific objectives of the supervisory development program are to:

- Demonstrate effectively that loss control is an integral part of, and depends on, overall operations control.
- Establish that improved operations lead to fewer losses.
- Emphasize that a supervisor is responsible for operations control and loss control in his department.
- Present a definite plan for investigating losses and improving operations, and to train supervisors in the use of this method.
- Afford a method of approaching the everyday supervisory responsibilities of training employees, improving job methods, and handling job relations problems as part of the supervisor's job of operations control and loss control; indoctrinate supervisors in how the use of these plans leads to improved supervisory performance.

The intial session is designed to create the workshop atmosphere. The workshop leader is expected to involve supervisory persons of similar status to acquire common interests. The primary purpose of the first session is to:

- Introduce the workshop program to supervisors as part of the management approach to loss control.
- Emphasize the relationship of loss control to management control in the department by pointing out the three factors affecting production: facilities, processes, and people.

- Identify the supervisor's role in recognizing and controlling conditions that can result in economic loss in the department.
- Present a uniform plan for identifying loss-producing conditions and train supervisors to apply corrective measures.

The assignment of various supervisors to participate as leaders during the subsequent workshop sessions should afford the opportunity for controlled progress and success of the workshop.

OUTLINE OF WORKSHOP SESSIONS
FOR DEVELOPING SUPERVISORY SKILLS

Session No. 1—Management Control for Loss Control

INTRODUCTION

The recognition of the conditions that lead to personal injury, damaged equipment and facilities, and loss of use of a facility is not difficult. The installation of appropriate control measures to reduce or eliminate the probability of loss does present a challenge to any activity. Every person of supervisory status must unite in an effort to implement the controls with some uniformity.

Every loss, regardless of the extent, indicates a breakdown in management control at the supervisory level most closely associated with the activity where the loss occurred. Losses resulting from similar conditions may not occur with sufficient frequency to alert the supervisor to the need for reinforcing his control. Also, the severity may vary from light to moderate without triggering an alarm that indicates something is wrong. However, a "once in a hundred years" type of occurrence resulting in severe loss may trigger an all-out effort to identify the conditions responsible and to install corrective measures.

I. DEFINITION OF LOSS CONTROL

The term "loss control" implies the control of conditions that can be responsible for a loss. The loss of property by fire or damage, or loss of personnel from injury has a definite influence on the financial status of any business activity. The majority of these losses are termed accidental. Let us briefly define the term accident: *An accident is the unintended interruption of the orderly progress to attain an objective.* Thus a loss is reflected in many ways, e.g. downtime, medical expense, and repairs, which are the direct yield of an accident.

The extent of dollar loss is not always recognized by management except when there is serious property damage or injury to personnel. However, we report and record certain related data to recover a portion of the loss through insurance. Some changes in job conditions and/or work practices may be instituted to correct the apparent cause of the accident that resulted in loss. There are some losses that appear to be uncontrollable in some activities every day. They are, more often than not, accepted as the normal course of events because the supervisor is not trained to recognize and control contributory conditions.

II. APPROACH TO LOSS PREVENTION

The projection of management's plan for loss control involves the fundamental application of some basic controls by supervisors. The primary factors involve *facilities, processes,* and *people*. It is recognized that executive policy related to these factors is already established to some degree. However, the promulgation of the policy at the activity level will afford the necessary controls to assure completion of any project in keeping with the objectives of the institution. Consideration is given here to each of these primary factors.

A. FACILITIES: *Specification, Arrangement, Occupancy, and Maintenance*
1. The facility manager's initial concern for facilities involves the site of operations. Here, he considers plant property and adjacent exposure, including traffic, buildings, utilities, the public, and employees. He selects specific areas of the initial operations to be compatible with these exposures.

    The specification for equipment for the initial and each subsequent phase of the operation is determined in accordance with a definite schedule for performing operations. However, size, capacity, economy of operation, and traffic planning are important for calculating costs.
2. The arrangement of equipment inside and outside of the facility should correspond to the flow of traffic, maintenance requirements, exposure to the elements, probability of fire, and other security exposures.
3. Supervisors who control the use and occupancy of equipment and facilities should be especially concerned with the performance of the employee. The acitivty supervisor should have standard operating practices and guidelines to measure the equipment's performance and predict its demand maintenance. These guidelines enable him to plan the activity for both reliability and prolonged productive life of the equipment.
4. The maintenance requirements for effective operation are usually outlined by the equipment manufacturer. To this end, a planned program for preventive and demand maintenance should be provided, with sufficient tools and qualified maintenance personnel available. Any abnormal frequency of demand maintenance should be sufficient cause for the unit supervisor to reevaluate the use of equipment.

B. PROCESSES: *Materials; Warehousing, and Operations*
1. The delivery of selected material to the activity site cannot always coincide with need. The availability of certain materials, source location, seasonal production, work stoppages, and

Fig. 1

volume of purchase may have considerable influence upon purchase and delivery. It is important to project these factors into the activity plan. Past performance of basic service materials can be considered when measuring costs, and should influence the selection of new materials for any similar subsequent project.
2. The warehousing of materials at the plant site should afford the least interference with other activity. The storage site should be accessible for delivery by the supplier to minimize liability exposure. Also, materials should be secure from theft, fire, and damage by environmental exposure.
3. The flow of materials in acceptable condition, from the storage area to the activity area, is a critical factor. Handling by mechanical means is economically feasible when the materials are protected against damage. Also, "strain and sprain exposure" of personnel is reduced when a specific procedure is outlined for moving materials.
4. The fabrication and assembly of materials in accordance with plans and specifications within acceptable tolerance levels is an important phase of any operation. The delay of schedule as a result of damaged material can be costly. Equally important

is the positive assurance that other materials and equipment will not be damaged in the activity processes. If the customer recognizes a defect or malfunction, in the equipment and material products after completion, the result is critical loss exposure, from a products or completed operations standpoint.

C. PEOPLE: *Recruiting, Assigning, and Leading*
   1. The commercial and industrial activities of today are given the opportunity to select people at top professional, technical, and semiskilled levels at the discretion of management. The application of "know-how" and the development of personnel for optimum performance appear to vary with the nature and size of the company. The full utilization of people who are members of trade unions is not always attained because of absence of effective management policy, or failure of supervisors to emphasize the policy and exercise the necessary control. To perform any function, people should be selected on the basis of specific qualifications, physical and mental ability, performance, and compatibility with other people.
   2. The training of a new or prospective employee starts with an introduction to management's written policy for loss control. This policy should include an introduction to the company's objectives and the role of the new employee. Initial orientation should familiarize the new employee with the total program, including reporting of accidents and the purpose of investigation of accidents; that is, the reduction of subsequent probability for loss.
   3. The assignment to a work position based upon qualifications should include an early supervisory evaluation of employee performance. The evaluation should consider attitude, attire, use of tools and equipment, actions with reference to life safety of self and others, and productivity in accordance with established standards.
   4. Identification of performance as good, standard, and substandard, without a supervisory plan for improvement, is probably the most serious deficiency in a program to control losses. The acceptance of less than desirable work practices by the supervisor is commonplace until serious injury or equipment damage occurs. The loss is then generally attributed to the poor performance of the employee, although in reality the failure of the supervisor's control was the condition most responsible for the loss.

A record of standard and substandard performance can be made and reviewed individually and collectively with the employee or craftsman. The standard procedure for a specific activity should be outlined. The activity supervisor cannot afford the luxury of assuming the employee "knows better" and is indifferent to the accepted practice. It is his responsibility as a representative of management to lead other supervisors and the work force to accomplish the objective of the company; that is, to *furnish all equipment, material, and people necessary to attain the company's objectives according to plans and specifications,* and with a reasonable profit.

III. CONCLUSION

The loss control effort in any enterprise cannot be limited to mere compliance with the minimum specification standards (job conditions) as developed by the occupational safety and health administration. The all-inclusive loss control program must include the necessary performance standards (work practices) and a definite plan of action to achieve this objective. It is here that management, the safety director, and personnel of supervisory status should cooperate to implement the specifics of the program: to plan, organize, and control for effective loss control.

Session No. 2—A Supervisor Investigates Conditions

INTRODUCTION

An investigation of conditions is an essential step in identifying the causes of accidents so that preventive measures can be applied. The risk management loss control program broadens this concept to include investigation of any situation or incident which interrupts or interferes with operations of an activity.

With an understanding of facilities, processes, and people—the basic tools of loss control—the supervisor learns how to explore conditions, in an orderly fashion, to identify necessary corrective measures. The supervisor investigates losses which have occurred in his own department. He prepares a report for discussion with other supervisors at conference sessions and for safety committee review. He discusses the findings on his own cases and the cases of other supervisors as well. Gradually he develops a definite pattern of thinking which helps him solve daily operating problems and improve his overall supervisory performance.

The supervisor, as a result of this training, learns to recognize in their early stages conditions which are causing—or can cause—losses. This is true loss control and is indicative of applied management controls.

I. POINT OF DEPARTURE

A. An investigation may be prompted by an injury to an employee, an incident which caused delays or damaged material, a loss of facility, or just some condition which is identified as being other than it should be. In any event, the investigative procedure is the same.

B. Progressive questioning gets answers. Use the "W" questions, WHY, WHAT, WHERE, WHEN, and WHO, as well as HOW. The order of questioning will sometimes vary, and backtracking may be necessary to resolve a condition that was not apparent at the start of the investigation. Determine *what happened* or *what is happening.* Then ask *"why?"* This helps decide *what should be done.*

30 / Loss Control in the OSHA Era

Fig. 2

II. IDENTIFYING THE RESPONSIBLE CONDITIONS

A. This is a guide to determine: *Why did it happen?*

IF APPARENT CAUSE APPEARS TO BE

Job Condition	Work Practice
What is apparent?	*Why* was it part of the job action?
Why wasn't it previously identified?	*Why* was it being done in this manner?
What contributed to its existence?	*What* was its purpose?
What was involved?	*What* details can be combined?
Where was the occurrence?	*What* details can be eliminated?
Where was its source?	*What* procedures were not followed?
Where can we have similar conditions?	*Where* can improvements be installed?
Where can we modify?	*Where* else are work practices influenced?
When did it occur?	*When* are the practices prevalent?
When do similar conditions occur?	*Who* was involved in the occurrence?
Who was responsible for it?	*Who* can demonstrate what was being done?
Who is involved in resolving it?	*Who* can provide answers?
How can it be corrected?	*How* will corrective measures affect others?
Who should apply corrective measures?	*How* can these measures be applied in other areas?
How can these measures be applied?	

III. WEIGH AND DECIDE

A. Get all the facts by studying the job and conditions where the loss occurred. The condition responsible for an accident will always involve one or more of the items related to the following:

Facilities	Processes	People
Selection of tools	Materials	Recruiting
Arrangement of equipment	Warehousing	Placement
Correct application	Handling	Training
Proper maintenance	Operations	Leading

"Losses Interfere With Our Company Objectives"
SUPERVISOR'S CONDITION REPORT

Activity or Employee's Name	Time of Occurrence	Date
Department	Function	

What Happened? — Describe the loss or condition that caused you to make this investigation.

Why Did It Happen? — Determine facts by studying job conditions and work practices. DETERMINE WHY – WHAT – WHERE – WHEN – WHO – HOW.

What Are The Control Improvements? — Determine which areas require additional attention.

FACILITIES	PROCESSES	PEOPLE
Equipment	Supplies	Recruit
Arrange	Storage	Assign
Use	Handling	Train
Maintain	Operation	Lead

What Have *YOU* Done Thus Far? — Install corrective measures to the extent of your responsibility. Follow up!

What Remains To Be Done? — OBJECTIVE . . . PROTECT OUR ASSETS!!

Supervisor	Date	Loss Control Coordinator	Date

Fig. 3

IV. CORRECTIVE MEASURES

A. The investigation of conditions responsible for a loss is productive only when there is realignment of conditions to eliminate or reduce the probability of similar losses. Too often the corrective measures are applied to the apparent cause and limited to the specific occurrence. Infrequently a supervisor will recognize the need to generally apply the corrective measures he has identified throughout the department. This may have an adverse effect unless the corrective measures apply specifically to the activity with similar conditions and there is sufficient instruction to get voluntary compliance from the entire staff.

B. The supervisor's investigation of a loss, or a condition where there is probability of loss, may develop some control measures that apply in other departments. Here, the value of the general safety committee with a loss control director is recognized. The committee with knowledge of all activity areas and the loss control representative with the knowledge of technical standards can develop meaningful Standard Procedure Instructions to reduce the probability of further loss.

Session No. 3—Skill in Planning

INTRODUCTION

The development and job progress of the supervisor depend on measured performance. The measurement is not random. Various attributes are considered, such as, neatness, congeniality, honesty, and general peer acceptance. While all are desirable traits, they do not in themselves frame the supervisory structure.

I. THE FIVE NEEDS OF A SUPERVISOR

A. The supervisor as a representative of management gets things done through other people. Therefore, within the management cycle he must plan, organize, and control conditions within the scope of his responsibilities toward attainment of the overall activity objective.

B. The initial phase of the management cycle is appropriately concerned with planning. Here one of the basic needs of a supervisor is to have knowledge of the work. We emphasize this as only one of the basic needs. Too often, in the past, some employees have failed miserably because promotions were made to supervisory positions solely on the basis of knowledge of the work and ability to perform within the scope of that knowledge. The other skills, such as planning, training, and leading, in addition to expanding the knowledge of responsibilities are necessary for the successful attainment of the activity objectives.

33 / The Educational Program

Fig. 4

II. PLANNING TO IMPROVE JOB METHODS
   A. Select a job activity.
      1. Select one that has or may produce losses or production problems.
      2. Select any other task at random to identify areas for improvement consideration.
   B. Break down the job.
      1. List all details of the job exactly as done by the present method.
   C. Examine every detail.
      1. Use these types of questions:
         WHY is it necessary?
         WHAT is its purpose?
         WHERE should it be done?
         WHEN should it be done?
         WHO is best qualified to do it?
         HOW is it best to do it?
      2. Also examine the following:
         Selection, arrangement, use, and maintenance of the involved *facilities*.
         Selection of materials, warehousing, handling, and operations related to the *process*.
         Recruiting, placement, training, and leading of *people*.

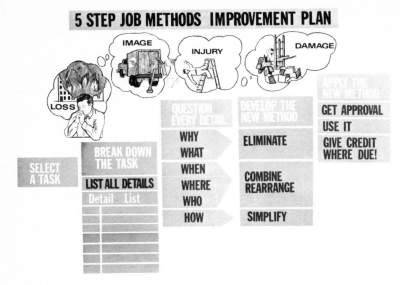

Fig. 5

D. Develop the new method.
   1. Eliminate unnecessary details; they are time-consuming and costly.
   2. Combine details when practicable.
   3. Rearrange for better sequence.
   4. Simplify all necessary details.
      Work out your ideas with others.
E. Apply the new method.
   1. Present your proposal to your supervisor.
   2. Get final approval from all concerned.
   3. Present the new method to the employees.
   4. Put the new method into operation.
   5. Give credit to all involved.

Session No. 4—Skill in Training

INTRODUCTION

The identification of good, standard, and substandard employee performance is the absence of a supervisory plan for improvement is probably the most serious deficiency in a program to control losses. The acceptance of less than desirable work practices by the supervisor is readily apparent when a loss occurs that results in serious employee injury or major equipment damage.

Training or job instruction is nearly always associated with the new employee on a new job. There is a wide variation in the training afforded to regular employees on new jobs. There also appears to be little or no effort to retrain old employees to improve work practices and increase yield.

I. THE WHO, WHEN, AND WHERE OF TRAINING
   A. One universal training timetable is sufficient to encompass the total training effort to be expanded for the new employee on the new job, the regular employee on a new job, or the old employee on the same job. The difference is at the point where the training is to start. There is some preparation necessary before any training session.
   B. It is noteworthy to emphasize the initial training for new employees which includes management's policy for loss control. The activity objectives and the role of the new employee should also be included. This employee orientation should detail the procedure for reporting any type of accident and explain the purpose of the supervisor's investigation.
   C. A record of standard and substandard performance can be made and reviewed individually and collectively with the employee. The standard procedure for a specific activity should be outlined. The activity supervisor cannot afford the luxury of assuming the employee "knows better" and was indifferent to the accepted practice. It is his responsibility as a representative of management to lead employees and other supervisors of the work force to attain the objectives of the activity.

II. PREPARATION FOR TRAINING
   A. Determine the extent of training to be provided and have a training timetable.
      1. Note the jobs.
      2. Decide who should be trained or retrained to do the jobs.
      3. Decide when they should be trained.
   B. Break down the job.
      1. List the important steps.
         An important step is a logical sequence of the operation when something happens to advance the work.
      2. Pick out the key points of each step.
         A key point is any important part of a step which will help prevent accidents and will make the work easier to do.
   C. Have everything ready.
      1. The job, including right equipment and material, should be available.
   D. Have the work place properly arranged.
      1. Show the employee how he will be expected to keep it.

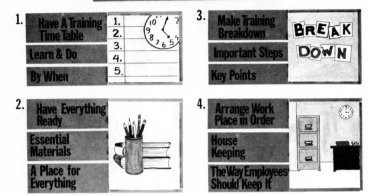

Lack of readiness...?

Fig. 6

III. PROGRESSIVE STEPS IN TRAINING
   A. Prepare the employee.
      1. Put him at ease.
      2. Start with what he knows.
      3. Get him interested in the job.
      4. Place him in the correct job position.
   B. Present the job.
      1. Tell, show, and illustrate carefully and patiently.
      2. Stress the key points—explain reasons.
      3. Instruct clearly and completely—one point at a time, no more than he can master.
      4. Repeat as necessary.
   C. Try out the performance.
      1. Have him do the job—correct errors.
      2. Have him do the job again, explaining key points.
      3. Ask "W" type questions.
      4. Continue to have him do the job until you know he knows it.
   D. Follow up.
      1. Put him on his own.
      2. Designate to whom he should go for help.
      3. Check frequently. Watch key points.

37 / The Educational Program

4. Encourage questions.
5. Taper off extra coaching and close follow-up.

If the worker hasn't learned, the instructor hasn't taught!

Session No. 5—Skill in Leading

INTRODUCTION

Loss control through better job relations can be considered to be an important factor in any activity's operation. Here, the supervisor, in his effort to get all his people to do their work in the manner in which it should be done and because they want to do it that way, can measure his skill in leading. There are some supervisors on each end of the norm— those with activity-wide recognition for getting work done on time and in quantity, but unpopular with the work force, and those recognized for getting work done with difficulty, but liked by their employees. Also, there are supervisors that get work done on schedule with even wear and tear on equipment, material, and people.

I. CONTROLLING ACTIONS

   A. A supervisor gets results through people.
      1. The results are measurable by attendance, quality, schedules, costs, production, accuracy, training, and service.

4 STEPS IN TRAINING

IF...

Fig. 7

B. The foundations of good job relations are:
   1. Let each person know how he is getting along.
   2. Give credit when due.
   3. Tell people in advance about changes that will affect them.
   4. Make best use of each person's ability.
C. People are a supervisor's raw material. They must be treated as individuals. Each has different health, finances, family, background, education, job, and attitude.

II. HOW TO HANDLE A JOB RELATIONS PROBLEM
   A. Get the facts.
      1. Review the record.
      2. Find out what rules and plant customs apply.
      3. Talk with individuals concerned.
      4. Get opinions and feelings.
      Be sure you have the whole story.
   B. Weigh and decide.
      1. Fit the facts together.
      2. What possible actions are there?
      3. Consider objectives and the effects on the individual, group, and production.
      Don't jump to conclusions.

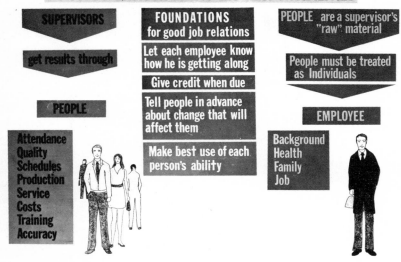

Fig. 8

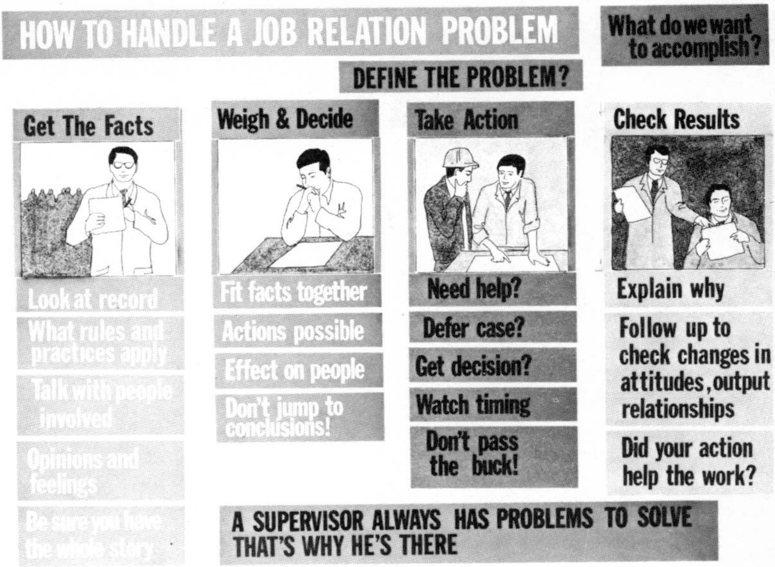

HOW TO HANDLE A JOB RELATION PROBLEM

What do we want to accomplish?

DEFINE THE PROBLEM?

Get The Facts
- Look at record
- What rules and practices apply
- Talk with people involved
- Opinions and feelings
- Be sure you have the whole story

Weigh & Decide
- Fit facts together
- Actions possible
- Effect on people
- Don't jump to conclusions!

Take Action
- Need help?
- Defer case?
- Get decision?
- Watch timing
- Don't pass the buck!

Check Results
- Explain why
- Follow up to check changes in attitudes, output relationships
- Did your action help the work?

A SUPERVISOR ALWAYS HAS PROBLEMS TO SOLVE THAT'S WHY HE'S THERE

Fig. 9

 C. Take action.
  1. Do you need help in handling?
  2. Should you defer action at present?
  3. Should you refer this to your supervisor?
  4. Watch the timing of your action.
 Don't pass the buck.
 D. Check results.
  1. Explain why action was taken.
  2. Watch for changes in output, attitudes, and relationships.
 Did your action help operations?

CHAPTER FOUR
The Engineering Program

The term "engineering" as related to the loss control program shall apply to the use of the arts and sciences in connection with facility design, equipment installation, and arrangement of tools and materials. The scope of engineering activity will encompass: (1) surveys and analysis, (2) research and development, (3) rearrangement and modification, and (4) engineering survey reports.

1. Surveys and Analysis

Surveys of facility installations by loss control department personnel shall be performed to measure effectiveness of program implementation by the various department heads and to reinforce the effort in this direction. The survey shall include a study of the operations to determine conditions responsible for losses.

The survey shall be conducted in an orderly manner. It will include the following points:

- *Statement of the problem*—presented by loss analysis or by recognizing conditions that can be responsible for loss.

- *Facts and observations*—disclosed by visiting and studying the operations in the field.
- *Conclusion*—a summary of facts and a projected view for recommendations after analysis.
- *Recommendations*—a realistic means of correcting conditions that are responsible for or could contribute to losses.

2. Research and Development

Whenever a major change is proposed in a general area of activity, the probability of loss as a result of that change shall be investigated at the earliest possible opportunity. The loss control department should be notified of contemplated changes in facilities and occupancy so that it can coordinate all necessary controls to reduce the probability of loss.

Research shall include loss control studies of new equipment and appurtenances which may be added to existing equipment to improve operations and reliability or prolong useful life and economical function. The development of new methods and procedures shall be given every consideration when a more productive yield and a curtailment of losses can be realized.

3. Rearrangement and Modification

While it is understood that the majority of losses involving injuries are the direct result of human behavior, it is also understood that certain changes can be made to minimize the accident and injury potential. These changes may be influenced by the degree of exposure and the possible improvement in operations that will be reflected in cost savings and productive yield.

The Supervisor's Investigation Report can be effectively utilized by department heads, supervisors, and loss control department personnel for the purpose of studying the operation to determine the advantage of rearrangement or modification and to determine whether or not losses are being experienced without being identified as such.

4. Engineering Survey Reports

The results of any survey and analysis should be prepared and presented in an orderly fashion. Each survey of an activity is performed with a specific objective, which is to reduce the probability of loss.

The loss control director or his representative, working with the risk manager, should already have a general knowledge of the "big picture" relating to the corporate and company function. They are familiar with facility locations, general arrangement of equipment, different processes, storage and stockpiling facilities, production flows, warehousing, and shipments. However, too few of these people are working objectively through surveys and analyses to enhance management's control of conditions that could have a definite influence on the corporate profit yield.

It is necessary to set objectives with priorities when scheduling surveys. Consideration should be given to the past performance, loss history, and readily identifiable degree of exposure. The schedule must be flexible enough to permit interruption for application of skills on a priority basis in certain other areas that cannot be delayed because they are not routine in nature.

We have provided an example of a survey of a medical facility that experienced a change in facility, function, occupancy, and people. The degree of exposure to loss was changed from that of a small company-owned hospital and clinic, treating company employees and supplying only emergency service for the public, to that of a general hospital for the public, with teaching facilities for the college of medicine at the adjacent university. The names and places are fictitious; however, the conditions are real, and the controls are considered necessary as part of the loss control effort for any similar medical facility.

It is desirable to include as many risk categories as possible in each loss control survey. However, the degree of exposure, loss experience, and extent of existing controls may permit separation of a particular category for future attention. Some additional risk categories for our hospital could involve future studies for money and securities, ambulance services, industrial injuries, a heliport, and a rehabilitation facility.

The loss control survey is not a prepackaged mix that can be put together with little effort. It is made from scratch using the basic risk categories as the prime ingredients. The refinements are added as they are recognized during preliminary work that includes library and reference research. The effectiveness of the loss control survey is measured in improved operations and reduction of exposure to loss.

A LOSS CONTROL SURVEY OF THE SAN PHELCOTT HOSPITAL
1850 Sentinel Drive, Temple Arizona

Prepared by The Risk Management Department
San Phelcott Copper Company, 3700 Commodore Avenue
Raines, New Mexico, October 1973

1.0 NATURE OF OPERATIONS

1.1 The San Phelcott Hospital at the Temple Medical Center at 1850 Sentinel Drive, Temple, Arizona, maintains operations as a privately endowed hospital with an outpatient clinic for company employees and the general public.

The hospital provides general, pediatric, and intensive patient care for communicable disease, mental illness, alcoholism, drug addiction, internal medicine, and corrective surgery. The facility is not involved in geriatrics or extended care activity, and there are no satellite activities operated by the hospital for these purposes or that of a sanitarium.

1.2 The facility maintains 250 patient beds, 20 bassinets and Isolettes to offer treatment and care for almost all medical specialties. The outpatient facilities provided examination and treatment for 29,000 outpatients from September 1972 through September 1973.

1.3 The hospital was initially designed and built to serve as a private facility. However, growth of the community and adjacent Marcus Del Sol University led to the construction of additional facilities during 1972 to include a teaching hospital for the college of medicine of the university which was to be an integral part of the Temple Medical Center. In this connection, the average number of "on duty" physicians, surgeons, interns, registered nurses, licensed practical nurses, aides, and orderlies is greater than in other community hospital facilities. It is also noteworthy to mention the total absence of claims by patients for alleged malpractice, errors, and/or occurrences involving out-of-the-ordinary exposure to injury or illness.

1.4 A genuine effort is made to obtain service personnel and technicians of the highest caliber. All are required to complete a written application for employment. Reference checks are made, and results are filed with the application. A preemployment physical examination is required for all employees, and an annual physical examination is performed. The initial physical examination for employees who were involved with ionizing radiation during previous employment does not include a blood test at the time of this survey.

1.5 There are written procedures for training domestic and physical resource employees to identify loss-producing conditions and to maintain acceptable job performance standards. Performance is reviewed on a regular schedule. New employees are reviewed more often, and additional instruction is provided if needed.

1.6 The training programs for fire and evacuation, disaster planning, disruptive disturbances, and bomb threats are presently being reviewed in the security section of the loss control department for possible refinement. Emergency organization drills are conducted every three months on all shifts. A full-time safety representative is assigned to the facility to coordinate the activity of the medical center personnel and the physical resources personnel for loss control. Also, specialists in radiation control, electrical apparatus, and security are involved in the loss control effort with specific duties in these areas.

2.0 ARRANGEMENT OF FACILITY

2.1 The San Phelcott Hospital of the Temple Medical Center is a nine-story building with a basement, reinforced concrete structural frame, floors and walls, identified as construction Class B. The exterior facing is mixed construction of mainly architectural precast concrete panels and brick facing with architectural coursing at the lower three levels. The initial construction was completed in 1961, and additional construction for initial occupancy as a teaching hospital was started about September 1, 1971. Less than 25 percent of the design features for the interior construction are combustible. The fabric-supported vinyl wall covering installed on interior partitions has a less than 25 percent flame spread.

2.2 Vertical openings include stairways, laundry chutes, and elevators. All are enclosed with noncombustible building materials with Underwriter's Laboratory (U.L.) approved protection at the openings. The discharge of the laundry chute can be modified to provide more secure closing of the bottom door. Laundry is done elsewhere by contractor.

2.3 Corridors are provided with electromagnetically controlled self-closing smoke stop doors. These are under consideration for replacement with approved U.L. fire-rated doors for increased protection. Emergency egress from each floor into Class A stairwells is available via four defined routes. Doors in these stairwells are U.L. approved with panic hardware.

2.4 The entire facility is provided with zone controlled air conditioning and heating. Hot and chilled water and processed water are supplied from a separate central plant building approximately 600 feet from the hospital building. There are three gas-fired steam boilers capable of 25,000 pounds per hour, and manned around the clock by stationary engineers. They are inspected daily by the facility engineers and semiannually by the national board licensed insurance inspectors.

2.5 The facility has three auxiliary natural-gas fueled generators for emergency electrical power and lighting. They have a capacity of 2,250 kilovoltamperes output to provide power and lighting for the operating rooms, intensive care units, emergency room, nursery, exit lights, stairway lights, and kitchen. The system is on the line with automatic start-up within seven seconds of commercial power failure. A manual override provides for operational testing once per week under load for a period of one hour.

2.6 Electrical switchgear, secondary power panels, distribution lighting panels, and equipment disconnects in accordance with the national and local code requirements. All wiring is in conduits with ground bushings at each device, and a full three-wire system is installed for a positive grounding circuit.

3.0 FIRE PROTECTION

3.1 Public fire protection is furnished by the city of Temple with a rating of fire protection Class 3. The fire stations which would provide early response are Station No. 9 at 3200 East Apache Drive, with a response time of three minutes, and Station No. 1 downtown, with a response time of five minutes. Accessible hydrants are within 100 feet of the perimeter of the building.

3.2 Private fire protection includes partial coverage with automatic sprinklers in the basement (85 percent), first floor (25 percent), second floor (10 percent), and third floor (10 percent). The system valving, pipe size, quality, and connections are in accordance with NFPA no. 13. The areas are sprinklered to provide increased life safety and protection. The extent of application with respect to the total area of the facility does not afford additional credit consideration. The sprinkler system must be inspected by regular flow tests. The results of the tests are recorded in a log at the riser site.

3.3 There is a good distribution (sixty total units) of first-aid extinguishers which are approved for A, B, and C type fires. These are inspected each month and serviced semiannually by personnel of the physical resources department.

3.4 Fire alarms include central station supervision of automatic sprinklers, ionization detectors in return air ducts, and mechanical equipment and manual trip stations in corridors. An enunciator panel is in the central telephone room at the hospital with an automatic relay to the fire department by a special line.

3.5 The kitchen facility on the second floor is not provided with automatic extinguishers at the hood over ranges and cooking equipment. The discharge duct is protected by metal mesh filters that are cleaned by the mechanical department. Filters should be cleaned at least once per month.

4.0 PROFESSIONAL EXPOSURES

4.1 There is a written policy regarding the type of patients admitted and the reasons for exclusion of any specific cases. Also, each patient is furnished with a patient's guide that outlines features of the hospital that are important to him. The basic controls for visiting hours, storage of valuables, nonprescription drugs, meals, and restriction of under-age visitors are laid down in the guide. The guide is designed not to look like a rule book. However, each desired control is sufficiently prominent and can be easily identified by the patient.

4.2 Patients are identified by wristbands and identification cards. Newborn infants are identified in accordance with the Hollister system, use of a bracelet with date of birth, number, and mother's name and number. The foot- and handprints are also used for identification.

4.3 A program exists for reporting any unusual occurrences on an incident report form. The reports are investigated by the department head. Any corrective measure is determined by the department head and the form delivered to the hospital administrator. An incident review committee is not maintained nor are the incidents summarized to identify possible problem areas and to afford direction for uniform control measures.

4.4 The nursing service for the hospital maintains primary guidelines for each level of the nursing staff. Nurses are specifically trained concerning the importance of having signed orders from the physician before treatment or medication. In any situation where the physician cannot be present to sign the order, two nurses must receive the call. Each nurse must validate the order by repeating it back to the physician in the presence of the other. The order is subsequently signed by the physician at the earliest possible time.

4.5 An in-service training program is maintained to update specific information as related to the most recent and accepted treatment methods. This training is also open to performance reviews and is intended to upgrade any less than desirable practices that could have adverse influence on the reputation of the institution.

4.6 Anesthetics are administered by an anesthesiologist. Anesthesia machines are grounded. These units are also maintained on schedule and tested for resistance to ground measurements. There is no scheduled culturing procedure on these units to monitor cleaning effectiveness.

4.7 There is at least one registered x-ray technician on duty at all times. All persons involved in activity where there is ionizing radiation exposure are required to wear a monitoring film badge.

4.8 Written regulations are maintained for the operation and control of laboratories. Complete records are kept of all tests and analyses.

All written orders and reports are signed and filed with the patient's record. A duplicate copy is kept in the laboratory. Instruments in this area are regularly inspected for calibration and tested to ensure reliability.

5.0 DRUGS AND NARCOTICS

5.1 Written guidelines are fully enforced for ordering and dispensing of all drugs. Orders for administering drugs are always in writing and signed by a physician. The drugs are administered only by registered nurses after the proper identification of the patient. The registered nurses use disposable needles to give injections and/or intravenous liquids. However, a specially designated medication nurse is not maintained. The needle punctures sustained by employees involved in trash disposal have caused studies to be conducted by the hazard control department. Some corrective application was accomplished. However, this activity should be reviewed at irregular intervals with respect to individual performance and compliance with disposal methods.

5.2 A full-time registered pharmacist is in charge of the pharmacy. Written procedures regulate the list of approved drugs and the policy regarding the use of research drugs. Also, there are automatic-stop orders for narcotics, antibiotics, sedatives, and anticoagulants. Treatments for longer periods must be signed by a physician.

5.3 All drug cabinets and storage areas are inspected on a regular schedule to remove outdated or deteriorating drugs. Labels are checked for legibility and rate of use to maintain a fresh supply with only necessary reserves. Drugs kept at the nurses' stations are in locked cabinets in full view, with a red warning light activated when the cabinet is open.

5.4 The unit dose system is maintained for approximately 95 percent of the drugs dispensed by the pharmacy. Packaging for each patient is clearly identified, and drugs are delivered by pharmacy technicians to minimize loss and/or delay.

6.0 BLOOD SERVICES

6.1 The blood bank maintained at the facility is approved by the American Association of Blood Banks. Written approvals for treatment as well as forms for rejection of treatment are required from the patient or a responsible member of the family.

6.2 The blood storage facility is provided with a high and low temperature alarm, and the electrical system is connected with the emergency power system.

6.3 A comparative analysis is performed on the patient's blood specimen and donated blood as well as major and minor cross matching as ordered by the patient's physician.

7.0 SPECIAL CONTROL AREAS

7.1 The operating rooms are constructed and maintained in accordance with existing codes and approved practices. All electrical fixtures and apparatus less than 5 feet above the floor line are explosion-proof. Conductive flooring is as approved; conductivity tests are performed each month, and records of resistance to ground are maintained. The highest recorded resistance to date is 250,000 ohms. Operating room equipment has conductive rubber tubing, casters, and wheels. These equipment items are also monitored on schedule for conductivity. Records on cards should be supplemented with an annual cumulative graph to reflect variation; these are given to the administration and housekeeping departments. The conductivity tests should reflect the resistance to ground between 25,000 ohms and 250,000 ohms as a desirable norm.

7.2 Medical gases involve the use of the pin index system on all anesthesia machines. However, less than 1 percent of the gases are flammable. These gases are stored in the exterior ventilated gas storage building and kept apart from other gases.

7.3 Operating room procedures include a record of physicians who are authorized to use the facilities and a list of restrictions that are placed upon the types of operating procedures.

7.4 There is a committee in charge of sterilization and infection control which is made up of members from the medical administration and nursing services in cooperation with community health organizations. Techniques and procedures are maintained for discovery of infections and tracing their sources, and records are maintained for studying sources of infection. In this connection monitoring of operating rooms, apparatus, and equipment should be recorded.

7.5 Operating room practices include a tri-count system for sponges and needles. The Plexiglas board, anesthesia record, and operating room forms are completed. Sponges are also tagged for X-ray tracing. An instrument count is made prior to any operation and before cleaning and sterilization repack.

7.6 Recovery rooms are readily accessible, constantly supervised, and equipped with facilities for emergency resuscitation, transfusions, and other emergencies. An alarm system is also maintained for prompt response of proper physicians to attend the emergency condition.

7.7 Patient rooms are arranged for optimum control from the nurses' stations. Furniture and fixtures correspond with current approved design for convenience and patient comfort. Beds are low profile and equipped with side rails. Windows are maintained in a closed position with tamper-resistant releases. Beds are a minimum of 3 feet from the exterior walls and windows are set back in the wall line providing nearly flush conditions at the interior to preclude use of the sill as a shelf or seat.

7.8 The hospital has organized committees that meet each month for administration, credentials, infection, audit, utilization review, tissue, forms, and safety. Also a building committee is maintained to review any proposed changes in arrangement, building facilities, or services.

8.0 CONCLUSIONS

8.1 The facilities of the San Phelcott Hospital of the Temple Medical Center, including administration and staff, present a very favorable resource for medical care for the sick and the injured. Also, programs designed for this teaching hospital to investigate, demonstrate, and promote means of achieving health care of the highest quality are indeed an asset to the medical profession.

8.2 The identification of every possible condition that may contribute to a loss is not possible during one survey period. Therefore, attention to the utilization of facilities, processes, and people by the loss control personnel employed at this facility is important to assist in the overall loss control effort. The basic structure, arrangement, and organization of the facility will remain essentially the same over a long period of time. However, the individual loss control studies of the various separate activities should relate to the exposures, identify the probability of loss, and offer proposals to control or eliminate the exposure to loss.

9.0 RECOMMENDATIONS

9.1 A corporate loss control manual should be furnished to the San Phelcott Hospital to provide a good base for persons with supervisory status. This should be a general program which includes education, engineering, and compliance for loss control. The policy statement, assignment of responsibility, and function of the safety committee should be part of, and give emphasis to, the manual. (See pages 13 and 14.)

The loss control manual is intended to be a reference guide for all staff members with supervisory status. Therefore, it should be couched in simple, easily understood language and have basic references numbered for use throughout the hospital.

9.2 The loss control representative should be involved in a continuing program to assist the heads of departments and supervisory persons in education and instruction for loss control. This extends beyond presentation of films, data sheets, and a review of existing procedures after a loss occurs. There should be loss control studies dealing with equipment, facilities, materials, and people in specific areas of the activity. The studies should involve the basic physical conditions and practices of individuals that conflict with standard procedures. They may be initiated because of an unusual frequency of occurrences, or merely because of the degree of hazard involved in an activity.

The OSHA standards, published by the Department of Labor as *Federal Register,* 29 CFR 1900, July 1973, vol. 37, no. 202, part II, are an excellent basic reference for determining areas for loss control studies. Also, additional standard procedure instructions can be developed from the appropriate subparts of the register.

9.3 In the investigation of losses, the primary concern should be to identify conditions that were responsible for the occurrence. The most interested person should be the supervisory person, since the occurrence reflects a breakdown in his effort to control the activity. However, the supervisor needs help in further developing his supervisory skills.

A standard investigation procedure should be established to involve the supervisor most closely associated with the activity where a loss has occurred. The written result of the investigation in simple form is not intended as an additional burden, but it will identify to the supervisor an area in need of improved supervisory control for optimum effectiveness. (See session no. 2 under the "Workshop Sessions for Developing Supervisory Skills" for a sample of an investigation report form.)

9.4 The hospital incident report should be considered to be privileged information as related to any unusual occurrence. The intent of the report is to identify a situation or condition which may involve a patient, visitor, vendor, or others, whether or not an injury or other illness was experienced. To maintain privileged information status, only the information concerning conditions responsible for the loss and the recommended corrective measures by the incident review committee should be included in a monthly summary for review.

9.5 The measurements of resistance to ground on conductive hospital equipment and floors should be recorded in graph form on a calendar year basis, for comparative analysis by the administration, housekeeping, and the hazard control department. This will permit early identification of an insulation trend and allow corrective application and/or scheduled replacement of any unreliable equipment.

9.6 The discharge door of the laundry chute should be modified to permit full function as an approved door to prevent propagation of a fire from one level to another.

9.7 The schedule for cleaning metal mesh filters should not exceed monthly frequency. Also, an automatic fire extinguishing system should be installed in the exhaust hood. The system should also have features for automatic shutdown of the heat source.

9.8 The culturing procedure should be restored to check cleaning effectiveness on anesthesia machines. Scheduled random tests conducted in operating, obstetrical, and recovery rooms should be recorded.

9.9 The initial preemployment physical examination for prospective employees who have worked with ionizing radiation during previous employment should include a blood test. The results should be recorded for comparison with any subsequently scheduled or interim tests.

CHAPTER FIVE

The Compliance Program

The term "compliance" is used to designate control measures established to ensure total performance of the program. The control measures usually are within the activities monitored by the loss control department. They are as follows: (1) rules and regulations, (2) standard procedure instructions, (3) traffic laws, (4) performance evaluation, (5) occupational safety and health standards, (6) investigation of losses, and (7) summary analyses.

1. Rules and Regulations

The rules and regulations governing the conduct of all personnel shall be in accordance with practices published, approved, and provided to all department heads and other designated persons of supervisory status. Regulations published and furnished to all persons operating company owned motor vehicles and mobile equipment are absolute requirements for continued employment.

2. Standard Procedure Instructions

These instructions are usually specific and relate to a particular activity. The instruction is often in written form to supervisory personnel. The degree of implementation is readily identified by the conditions maintained in the activity and/or the accident experience. The instructions provide guidelines of what to *do* rather than what *not to do*.

3. Traffic Laws

Regulatory laws established by state and local authorities shall be made available to all employee operators concerning their area of activity. These laws will be periodically reviewed with vehicle operators through defensive driving courses and other instructional media.

4. Performance Evaluation

The effectiveness of the loss control program will be determined and reported to executive management annually. Interim evaluations will be made by the loss control administrator on a quarterly schedule to identify areas in need of reinforcement. The evaluations may include survey reports and recommendations of insurance company consultants or other consulting firms approved for this purpose.

All areas of activity shall be monitored for compliance with the program. The loss control department shall conduct reviews of operational activity and consult with department heads to determine possible reinforcement areas for its loss control effort and to decide on corrective measures necessary to properly implement the program in that area of activity.

5. Occupational Safety and Health Standards

The published standards in the *Federal Register* for occupational safety and health are the law and are to be considered as supplemental to your loss control manual. Each department head is responsible for incorporating the standards outlined in the loss control manual into his operations to control physical conditions and work practices.

To assist the department head in his managerial responsibilities, specialists from the loss control department will analyze unusual activities to identify areas suspected of violation. Areas involving environmental controls such as toxics, flammables, dusts, and noise may require testing and documentation beyond your capacity. These should be identified at the earliest date, and a consulting service should be contacted after management gives approval. Subsequent controls shall be maintained as a supervisory responsibility.

6. Investigation of Losses

When investigating losses, the loss control department should be primarily concerned with identifying conditions that were responsible for the occurrence. The most interested person should be the supervisor, since the occurrence reflects a breakdown in his effort to control the activity. However, the supervisor needs help to further develop his supervisory skills, and the loss control department is in the best position to provide the help as a result of the investigation.

A standard investigation procedure shall be established to involve the supervisor most closely associated with the activity where a loss has occurred. The written report of the investigation in simple form is not intended as an additional burden; however, it will identify to the supervisor an activity in need of supervisory control for optimum effectiveness. A sample investigation report form is shown in session no. 2 under the "Workshop Sessions for Developing Supervisory Skills."

7. Summary Analyses

To determine the adequacy of the overall management effort, each quarter, executive management shall be provided with a summary analysis of reported losses. These summary reports shall reflect:

- The total number of first-aid injuries, classified by activity and location, type of injury, and the nature of the activity.
- The total number of disabling injuries and injuries requiring professional medical treatment, classified by activity and location, type and source of injury, and the nature of the activity. Each satellite location will complete and post OSHA Form 102, as required, annually.
- The total number of motor vehicle accidents involving injury, cargo loss, and property damage, classified by activity area and estimated amount of dollar loss.
- The total number of occurrences involving property damage,

such as fire, wind, water, vandalism, etc. The estimated dollar loss and appropriate commentary should be included in a report to the safety committee for review of corrective measures and control possibilities in this area.

The summary report shall be reviewed by supervisory persons, loss control department personnel, and others, including loss control consultants. They will determine the adequacy of applied corrective measures, and propose any methods for improvement. The summary report and the Supervisor's Investigation Report will identify to the loss control administrator and the department heads areas where supplemental work is required to reinforce corrective measures or to modify and rearrange facilities, equipment, and personnel. For examples of the summary forms, see "The Summary of Loss Experience" section of this book (pp. 61-70). The forms are adaptable and can be tailored to meet almost any requirement for measuring the loss experience.

THE PERFORMANCE EVALUATION PROGRAM

The effectiveness of management controls for loss control can be readily measured by using this basic system. To provide an accurate base for measuring the progress of subsequent management and staff effort, the measurement should be recorded at the earliest opportunity after the completed initial survey of a facility. Also, the continued measurement of losses should reflect a decrease in frequency and dollar loss to correspond with the continued improvement in measured performance.

There are several advantages associated with an assessed and recorded measurement of the loss control effort. It provides a vehicle to convey to management in understandable, concrete terms the status of the loss control effort. Also, the measurement "score" is easily understood.

An assessed and recorded measurement will bring into proper perspective the various applied loss control measures, on a continuing basis. This enables the loss control coordinator to promote through the facility's management, in an understandable way, those loss control measures which are considered to be the most effective.

The evaluation process incorporated in the technique requires a degree of guided self-evaluation. The local management must examine its own operation under the guidance of the evaluator to determine the status of loss control activities. The areas that reflect less than desirable controls are readily identified. The loss control coordinator and the department head will have some indication of which way to direct their efforts for improvement in specific areas. Also, the planning effort for setting objectives is furnished with some substance for consideration.

The evaluation process assists the facility's management to become familiar with corporate objectives, and one hopes they can be motivated to strengthen, as part of a short-range program, those areas seriously in need of improvement.

This type of measurement provides a basis for comparative evaluation. A comparison of the scores from one time period to the next is valid to determine progress. Within a multilocation activity, a comparison of the various locations can be made.

The evaluation, needless to say, should be guided by a safety professional who has a good knowledge of the operations and the expected loss control performance.

PERFORMANCE EVALUATION
RATING FORM

ACTIVITY:
LOCATION:

A. MANAGEMENT CONTROLS

	Poor	Fair	Good	Excellent	Comments
1. Statement of policy and assignment of responsibilities	0	5	15	20	
2. Standard Procedure Instructions	0	2	15	17	
3. Employee selection and placement	0	2	10	12	
4. Emergency reporting procedures	0	5	15	18	
5. Executive management involvement	0	10	20	25	
6. Rules for specific hazard areas	0	2	5	8	
Category rating	___ + ___ + ___ + ___ × 0.20 rating ___				

B. SUPERVISORY SKILLS FOR MOTIVATION AND TRAINING

	Poor	Fair	Good	Excellent	Comments
1. Supervisor loss control training	0	10	22	25	
2. Orientation for new employees	0	1	5	10	
3. Job hazard analysis	0	2	8	10	
4. Training for specialized operations	0	2	7	10	
5. Internal self-inspection	0	5	14	15	
6. Safety promotion and publicity	0	1	4	5	
7. Employee/supervisor safety contact and communication	0	5	20	25	
Category rating	___ + ___ + ___ + ___ × 0.20 rating ___				

57 / The Compliance Program

		Poor	Fair	Good	Excellent	Comments
C.	LOSS REPORTING, RECORDING, INVESTIGATION, AND ANALYSIS					
	1. Loss investigation by supervisors	0	10	24	30	
	2. Loss analysis and summary report	0	7	24	30	
	3. Investigation of property losses	0	10	24	30	
	4. Proper reporting of losses and contact with insurance carrier	0	3	8	10	
	Category rating	____ +	____ +	____ +	____ × 0.20 rating ____	
D.	OCCUPATIONAL SAFETY AND HEALTH					
	1. Ventilation—fumes, smoke, and dust control and personal hygiene	0	5	18	20	
	2. Machine guarding, drives, and points of operation	0	5	16	20	
	3. General area guarding	0	5	16	20	
	4. Maintenance of equipment, guards, hand tools, etc.	0	5	16	20	
	5. Material handling—manual and mechanized	0	3	8	10	
	6. Safety appliances and apparel	0	3	6	10	
	Category rating	____ +	____ +	____ +	____ × 0.20 rating ____	
E.	FIRE PREVENTION AND ENVIRONMENTAL CONTROLS					
	1. Chemical hazard control references	0	6	17	20	
	2. Flammable and explosive materials control	0	6	17	20	
	3. Housekeeping—storage of materials	0	5	18	25	
	4. Fire prevention measures	0	2	8	10	
	5. Waste disposal, trash collection, and environmental controls	0	7	20	25	
	Category rating	____ +	____ +	____ +	____ × 0.20 rating ____	

SUMMARY

The numerical values below are the weighted ratings calculated on rating sheets. The total becomes the overall score for the location.

A. Management controls _____
B. Supervisory skills for motivation and training _____
C. Loss reporting, recording, investigation, and analysis _____
D. Occupational safety and health _____
E. Fire prevention and environmental controls _____

TOTAL RATING _____

RATING FORM GUIDELINE

A. MANAGEMENT CONTROLS

Poor	Fair	Good	Excellent
1. *Statement of policy and assignment of responsibilities*			
No written loss control policy. Responsibility and accountability not assigned.	A general understanding of loss control, responsibilities, and accountability, but not written.	Loss control policy and responsibilities written and furnished to supervisors.	Loss control policy is reviewed annually and is posted. Responsibility and accountability are emphasized in supervisory performance evaluations.

58 / Loss Control in the OSHA Era

Poor	Fair	Good	Excellent

2. Standard Procedure Instructions (SPIs)

Poor	Fair	Good	Excellent
No written SPIs.	Written SPIs for some but not all operations.	Written SPIs for all operations.	All operations covered by procedures and furnished to each supervisor. Also annual review to determine adequacy of each procedure.

3. Employee selection and placement

Poor	Fair	Good	Excellent
Preemployment physical examination given for some jobs only.	Also a qualification test is administered to new employees.	Also the new employees' past safety records are considered in their employment. Physical exams given for all jobs.	Also when employees are considered for promotion, safety performance is considered.

4. Emergency reporting procedure

Poor	Fair	Good	Excellent
No plan or procedures.	Intermittent memos for emergency procedures.	Written plan outlining the minimum requirements.	All types of emergencies covered with written procedures. Responsibilities are defined with backup personnel provisions.

5. Executive management involvement

Poor	Fair	Good	Excellent
No measurable activity.	Follow-up on loss problems.	Executive safety committee reviews all loss reports, and management holds supervisors accountable for installing corrective measures.	Also reviews all Supervisors' Investigation Reports. Loss control problems are treated as other operational problems in staff meetings.

6. Rules for specific hazard areas

Poor	Fair	Good	Excellent
No written rules.	Specific hazard rules have been developed and posted.	Specific hazard rules are incorporated in all loss control guidelines for supervisors' reference.	Also, specific hazard rules are firmly enforced and updated at least annually.

B. SUPERVISORY SKILLS FOR MOTIVATION AND TRAINING

1. Supervisor's level before receiving loss control training

Poor	Fair	Good	Excellent
All supervisors have not received basic loss control training.	All shop supervisors have received some loss control training.	All supervisors participate in loss control training sessions at least semiannually.	Also have specialized sessions conducted on specific conditions since previous training was completed.

2. Orientation for new employees

Poor	Fair	Good	Excellent
No guidelines for the health and safety job requirements.	Verbal only.	A written handout to assist in orientation.	A formal orientation program for new employees is maintained.

59 / The Compliance Program

	Poor	Fair	Good	Excellent
3. Job hazard analysis	No written program.	Job hazard analysis program performed on some jobs.	Job hazard analysis conducted on the majority of operations.	Also, job hazard analysis performed on a regular basis and Standard Procedure Instructions furnished for all operations.
4. Training for specialized operations (equipment, machinery, flammables)	Very limited training given for specialized operations for new people.	An occasional training program given for specialized operations and sporadic retraining.	Training is given for all specialized operations on a regular basis, and retraining given periodically to review standard procedures.	Also, an evaluation is performed annually to determine training needs.
5. Internal self-inspection	No written program to identify and evaluate substandard practices and/or conditions.	Facility relies on outside sources, i.e., insurance safety engineer, and assumes each supervisor inspects his area.	Written inspection guidelines, responsibilities, frequency, and follow-up.	Inspection program is measured by results, i.e., reduction in losses and costs.
6. Safety promotion and publicity	Bulletin boards and posters are considered the primary means for safety promotion.	Occasionally safety displays, demonstrations, and films are used.	Safety displays and demonstrations are used on a regular basis.	Special articles and displays are provided. Displays are used regularly and are keyed to a theme.
7. Employee/supervisor safety contact and communication	Supervisors infrequently discuss safety with employees.	Limited safety discussions between supervisors and employees.	Supervisors regularly cover safety when reviewing work practices with individual employees.	Also, supervisors make good use of the SPIs and regularly review job safety requirements with employees.

C. LOSS REPORTING, RECORDING, INVESTIGATION, AND ANALYSIS

	Poor	Fair	Good	Excellent
1. Loss investigation by supervisors	No investigation made by supervisors.	Supervisors investigate only disabling injuries.	Supervisors trained and make complete and effective investigations of all losses; corrective measures installed.	Every loss is investigated within 24 hours of occurrence. Corrective application is thorough.

Poor	Fair	Good	Excellent

2. Loss analysis and summary report

No analysis of disabling and medical cases to identify prevalent causes.	Effective analysis of both cause and location maintained on medical and first-aid cases.	Accident analysis; results are used to pinpoint losses.	Losses are graphically illustrated to develop the trends and evaluate performance.

3. Investigation of property losses

No program.	Management inquires about property damage accidents.	A requirement that all property losses be investigated.	Also, management requires an outline of corrective measures on all property loss and progress.

4. Proper reporting of losses and contact with insurance carrier

Loss reporting procedures are inadequate.	Losses are correctly reported on a timely basis.	Loss records are maintained for analysis purposes.	Also, there is a close liaison with the insurance carrier.

D. OCCUPATIONAL SAFETY AND HEALTH

1. Ventilation—fumes, smoke, and dust control

Ventilation rates are below industrial hygiene standards.	Ventilation rates in exposure areas meet minimum standards.	Ventilation rates are periodically measured, recorded, and maintained at approved levels.	Equipment is properly selected and maintained close to maximum efficiency.

2. Machine guarding, drives, and points of operation

Some attempt is made to guard machine drives.	Partial controls in evidence for point of operation guards.	Controls meet applicable federal and state requirements.	Machinery is effectively controlled to the extent that injury is unlikely.

3. General area guarding

No attempt is made to control walking and working surfaces.	Some random effort to control these hazards is evidenced.	Evidence that controls meet applicable federal and state requirements—can be improved.	These hazards are effectively controlled to the extent that injury is unlikely.

4. Maintenance of equipment, guards, hand tools, etc.

None	Some, but inadequate or ineffective maintenance.	A regular program for equipment and safety features.	Preventive maintenance is programmed for hazardous equipment and devices.

5. Material handling—manual and mechanized

No attempt is made to minimize losses.	Some attempt to control material handling losses.	Loads are limited in size and shape for manual handling. Equipment is provided for heavy or bulky loads.	Also, there is orderly flow of materials in the facility.

	Poor	Fair	Good	Excellent

6. *Safety appliances and apparel*

Poor	Fair	Good	Excellent
Proper equipment not always provided.	Some supervision for distribution and use of personal protective equipment.	Proper equipment is provided. Equipment identified for special hazards.	Equipment provided complies with standards. Supervision is maintained.

E. FIRE PREVENTION AND ENVIRONMENTAL CONTROLS

1. *Chemical hazard control references*

Poor	Fair	Good	Excellent
No reference library.	References used by supervisors on occasion.	Additional standards maintained and used.	Reference standards have been promulgated, reviewed with employees, and maintained.

2. *Flammable and explosive materials control*

Poor	Fair	Good	Excellent
Storage and handling facilities do not meet fire regulations.	Some storage facilities meet minimum fire regulations. Most containers labeled, and provided with approved dispensing equipment.	Also, facilities meet minimum fire regulations.	Storage facilities exceed the minimum fire regulations. Good control of the handling, storage, and use of flammable materials.

3. *Housekeeping—storage of materials, etc.*

Poor	Fair	Good	Excellent
Housekeeping is generally poor.	Housekeeping is fair. Some attempts to adequately store materials are being made.	Housekeeping and storage of materials are orderly.	Housekeeping and storage of materials are ideally controlled for ease of handling and fire prevention.

4. *Fire prevention measures*

Poor	Fair	Good	Excellent
Do not meet minimum insurance or municipal requirements.	Meet minimum requirements.	First aid is provided. Extinguishers on all welding carts.	A fire crew is organized and trained in emergency procedures.

5. *Waste disposal, collection, and environmental controls*

Poor	Fair	Good	Excellent
Very few control measures.	Some controls exist for disposal of harmful wastes or trash. Procedures exist for collection and disposal.	Most waste disposal problems have been identified and control programs instituted.	Waste disposal hazards are effectively controlled. Air/water pollution potential is minimal.

THE SUMMARY OF LOSS EXPERIENCE

Experience Records and Directional Guidelines

1. We have indicated that management's effort for loss control involves the total exposure of all corporate activities. Therefore,

it is necessary to record and analyze the losses before installation of immediate and long-range remedial measures.

2. Executive management should have a summary of loss experience at least semiannually. However, the frequency of the summary will vary with the size and complexity of the corporate activity. The management persons with the assigned responsibility, using the system outlined here, will have firsthand knowledge of loss trends. Also, the corrective measures installed in the interim can be reported to management in the summary analysis.

3. Consideration for summary analysis should be given to losses resulting from occupational injuries and illnesses, general liability claims (personal injury and property damage to others), automobile accidents, and property (real and portable). Examples are given as a simplified guideline. Once the format is established, it is a simple task to identify, categorize, and record each occurrence in a matter of minutes as a daily routine. The monthly, quarterly, semiannual, and annual summary can be organized with little additional effort. The examples presented, while fictitious, are basic worksheets for the loss control coordinator and the plant safety committee.

Measuring the Effectiveness of the Loss Control Program

1. The identification of all losses and the recording of the factual data will offer management an additional measurement of supervisory performance. The common indices describing the performance are outlined here:

   a. DISABLING INJURY FREQUENCY RATE — The number of disabling injuries per million employee-hours of exposure:

   $$\text{DIFR} = \frac{\text{disabling injuries} \times 1{,}000{,}000}{\text{employee-hours of exposure}}$$

   b. DISABLING INJURY SEVERITY RATE — The total number of days charged per million employee-hours of exposure:

   $$\text{DISR} = \frac{\text{total days charged} \times 1{,}000{,}000}{\text{employee-hours of exposure}}$$

   c. DISABLING INJURY INDEX — An index computed by multiplying the disabling injury frequency rate by the disabling injury severity rate and dividing the product by 1,000:

$$\text{DII} = \frac{\text{DIFR} \times \text{DISR}}{1{,}000}$$

    d. MOTOR VEHICLE ACCIDENT RATE — The total number of accidents, disregarding preventability, per million miles of operating exposure:

$$\text{MVAR} = \frac{\text{number of accidents} \times 1{,}000{,}000}{\text{miles of operating exposure}}$$

2. The loss rate is essentially the number of accidents for a given activity over a period of time distance and can be coupled with the dollars expended to reflect total loss for any given period as well as to identify the average cost per accident for any increment in the total time period.

An example is:

    a. DOLLAR LOSS RATE — The cost of the total number of accidents per million employee-hours of exposure or million miles of operating exposure:

$$\text{DLR} = \frac{\text{loss expenditure} \times 1{,}000{,}000}{\text{employee-hours of exposure or miles of operating exposure}}$$

3. The guideline usually employed by property and casualty insurance underwriters to measure performance is related to pure experience. This is identified as a loss ratio (LR), which is a comparison of direct dollar loss paid out in claims on behalf of the insured as compared to the premium paid to the insurance company.

An example is:

$$\text{LR} = \frac{\text{insurance company's loss payout}}{\text{paid premiums by the insured}}$$

The variation of loss ratio in the different lines of insurance coverages offers some latitude to the insurance underwriter. However, a constant loss ratio of 50 to 60 percent in all lines is indicative of poor management control. The loss experience gauged over a period of years will reflect the risk status and the extent of loss control service necessary to attain the status that will prove profitable for both the insurance company and the insured.

SUMMARY OF OCCUPATIONAL INJURIES AND ILLNESSES

Colorscope Incorporated, Permafrost, Arizona
January 1, 1973 through June 30, 1973

COLORSCOPE INCORPORATED
PERMAFROST, ARIZONA
OCCUPATIONAL INJURIES AND ILLNESSES

1973 Report Period 1-1-73 to 7-1-73

DEPARTMENT	Eyes-Foreign Body	Eye Burns-Flashes	Wounds-Puncture	Wounds-Lacerations	Burns-Heat	Burns-Chemical	Dermatitis	Fractures	Contusions	Slips and Falls	Strains	Sprains	Animal Bites	Respiratory	Poisons	Miscellaneous	Total Accidents	Total Man-Hours Lost
Administration																	0	
Advertising																1	1	
Art																	0	
Chemical						2	3										5	
Construction	3			3		1*											7	32
Engineering																	0	
Maintenance			2														2	
Manufacturing	4																4	
Quality Control				2													2	
Receiving																	0	
Sales																	0	
Security									1								1	
Service									2								2	
Shipping	6								1	1	3	1					12	
Warehouse	2									1		2					5	
Total Accidents by Type	15	2	2	3	0	2	4	0	0	2	4	5	1	1	0	0	0	41
% of Total Accidents	37	05	05	07	0	05	10	0	0	05	10	12	02	02	0	0	0	100%	32

Total Employees	750
% Employees Injured	5%
No. of First-Aid	36
No. of Dr. Cases	4
No. Disabling Injuries	1
Total Accidents	41

Remarks:
1. Disabling injury involved first degree burns to arms and neck of employee handling cement in bags is indicated by (1*).
2. Eye injuries were from blowing dust in March. Dock enclosure and windshield are now installed.
3. Sprains and strains contributed to four doctor cases. Additional training is started in Warehouse and Shipping Departments.
4. The disabling injury frequency rate based on 720,000 man-hours worked is

$$\text{DIFR} = \frac{1 \times 1,000,000}{720,000}$$ is 1.38 disabling injuries per million man-hours of work

Loss ratio = $5,350 accident costs ÷ $37,800 premium = 14%.

_____   _____   _____   _____
Date   Loss Control Director   Date   Chairman Safety Committee

Fig. 10

COLORSCOPE INCORPORATED
PERMAFROST, ARIZONA
WORK SOURCE ANALYSIS

1973
Report Period 1-1-73 to 7-1-73

DEPARTMENT	Falls (Same Level)	Falls (Diff. Level)	Point of Operation	Improper Storage	Machine Drives	Housekeeping	Environmental Conditions	Striking Against	Vehicles	Struck By	Hot Substance	Corrosive Substance	Miscellaneous	Total No. Accidents	% of Total Accidents
Administration														0	0
Advertising													1	02	1
Art														0	0
Chemical										2		3		12	5
Construction				3			3					1		18	7
Engineering														0	0
Maintenance							2							05	2
Manufacturing							4							10	4
Quality Control										2				05	2
Receiving														0	0
Sales														0	0
Security			1											02	1
Service	1		1											05	2
Shipping				1		4	6	1						29	12
Warehouse					2		2	1						12	5
Total No. of Occurrences	1	3	3	6	0	0	17	4	0	2	0	4	1	41	
% of Total Occurrences	02	07	07	15	0	0	42	10	0	05	0	10	02	100	

REMARKS:
1. Environmental conditions contributed most to the loss experience.
2. Placement of materials (improper storage) on docks and in warehouse contributed to strain-type injuries.
3. Supervisory effort should be reinforced in the areas of safety appliances and apparel, material handling and employee work practices (training).

Total Accidents: 41

_____   _____   _____   _____
Date   Loss Control Director Date   Chairman Safety Committee

Fig. 11

COLORSCOPE INCORPORATED
PERMAFROST, ARIZONA
PARTS OF BODY INJURED ANALYSIS

1973
Report Period
1-1-73 to 7-1-73

DEPARTMENT	Eyes	Head	Trunk	Arms	Hands	Fingers	Legs	Feet	Toes	Miscellaneous					Man-Hours Lost	Total Accidents
Administration																0
Advertising					1											1
Art																0
Chemical				2	2	1										5
Construction	3			1*	1	2									*32	7
Engineering																0
Maintenance	2															2
Manufacturing	4															4
Quality Control					2											2
Receiving																0
Sales																0
Security				1												1
Service				1	1											2
Shipping	6		2	1	1		1			1						12
Warehouse	2		1	1						1						5
Totals	17	0	3	7	8	3	1	0	2							41
% of Total	41	0	07	18	20	07	02	0	05						*32	
Total Accidents		41														

Remarks:
1. Eye injuries accounted for 41% of the loss experience.
2. Hands accounted for 20% and arms 18% of total injuries reflect employee work practices.
3. The hand injury charged to advertising department was the result of a bite from a burro during an advertising campaign.
4. Protection of parts of body exposed to the injuries as recorded can reduce the frequency.

Date   Loss Control Director   Date   Chairman Safety Committee

Fig. 12

COLORSCOPE INCORPORATED
PERMAFROST, ARIZONA
EMPLOYEE ACTIVITY ANALYSIS

1973
Report Period
1-1-73 to 7-1-73

DEPARTMENT	Hand. of Materials	Lifting	Use of Tools/Equip.	Inattention	Unsafe Positions	Cleaning, Oiling	Personal Hygiene	Lack of Safety Equip.	Jumping/Running	Climbing	Walking	Miscellaneous	Total No. Accidents
Administration													0
Advertising					1								1
Art													0
Chemical					3			2					5
Construction			2	1			1*	3					7
Engineering													0
Maintenance								2					2
Manufacturing								4					4
Quality Control	2												2
Receiving													0
Sales													0
Security										1			1
Service				1	1								2
Shipping	2	2		1				6	1				12
Warehouse	1	2						2					5
Total by Type	5	4	2	3	5	0	1	19	1	1	0	0	41
% of Total	12	10	05	08	12	0	02	47	02	02	0	0	

Total Accidents: 41

Remarks:
1. Employees not wearing safety appliances and apparel was responsible for 47% of the controllable activity.
2. Dock modification improved environmental condition to reduce blowing dust, however, availability of eyewear is necessary for unusual conditions.
3. *Handling of materials, manually and mechanically should have "Job Hazard Analysis."
4. The five minute loss control instruction periods for the peak activity periods should emphasize safety equipment and lifting and carrying practices.

_____ Date _____ Loss Control Director _____ Date _____ Chairman Safety Committee

Fig. 13

SUMMARY OF GENERAL LIABILITY CLAIMS
FOR ILLNESSES AND INJURIES
AND PROPERTY DAMAGE

Colorscope Incorporated, Permafrost, Arizona
January 1, 1973 through June 30, 1973

COLORSCOPE INCORPORATED
PERMAFROST, ARIZONA
GENERAL LIABILITY OCCURRENCES INVOLVING OTHERS (NONEMPLOYEES)

1973 Report Period 1-1-73 to 7-1-73

DEPARTMENT	Falls (Same Level)	Falls (Diff. Level)	Point of Operation	Improper Storage	Machine Drives	Housekeeping	Environmental Conditions	Striking Against	Mobile Equipment	Struck By	Hot Substance	Corrosive Substance	Products	Miscellaneous	% of Total Accidents	Total No. Accidents
Administration	1														16.7	1
Advertising	1	1													33.2	2
Art																
Chemical				1											16.7	1
Construction																
Engineering																
Maintenance			1												16.7	1
Manufacturing																
Quality Control																
Receiving																
Sales																
Security																
Service																
Shipping																
Warehouse						1									16.7	1
Total Accidents by Type	2	1	1	1		1										6
% of Total Accidents	33	17	17	17		17									100%	

Total Accidents	6
Claim Costs	$9,300
Insurance Premiums This Period	$27,000
Loss Ratio	34%

Remarks:
Loss ratio = $9,300 losses ÷ $27,000 premium = 34%
Two falls on same level were on terrazo floor area of offices.
One fall from different level was trespasser fall from platform of billboard.
One point of operation was finger amputation of outside company service man.
One improper storage and fire damaged adjacent property.
One customer slipped on oil in aisle of warehouse.

_____ _____ _____ _____
Date   Loss Control Director Date   Chairman Safety Committee

Fig. 14

SUMMARY OF AUTOMOBILE ACCIDENTS
Colorscope Incorporated, Permafrost, Arizona
January 1, 1973 through June 30, 1973

COLORSCOPE INCORPORATED
PERMAFROST, ARIZONA
SUMMARY OF AUTOMOBILE ACCIDENTS

1973
Report Period 1-1-73 to 7-1-73

DEPARTMENT	TWO VEHICLES						ONE VEHICLE									Nonpreventable	Preventable	Total Accidents
	Intersection	Vehicle Ahead	Vehicle Behind	Passing	Being Passed	Oncoming	Fixed Object	Pedestrian	Private Property	Passenger	Noncollision	Miscellaneous	Vacant Space	Parked	Backing			
Administration	1																1	1
Advertising																		
Art																		
Chemical														1			1	1
Construction				1			1								1	1	1	2
Engineering							1								1			1
Maintenance																		
Manufacturing																		
Quality Control		1															1	1
Receiving																		
Sales	1	1					1								1	1	2	3
Security																		
Service		1															1	1
Shipping				1	1												2	2
Warehouse																		
Total by Accident Type	2	3		1	2		3							1		3	9	12

Total Accidents	12
Total Direct Cost	$18,000
Miles of Operation	1,100,000
Loss Ratio	55%

$$\text{MVAR} = \frac{12 \times 1{,}000{,}000}{1{,}100{,}000} = 10.91 \text{ (number of accidents per million miles of operation)}$$

$$\text{DLR} = \frac{18{,}000 \times 1{,}000{,}000}{1{,}100{,}000} = \$16{,}363.63 \text{ (direct dollar loss rate per million miles of operation)}$$

Remarks:
Colliding with the vehicle ahead reflects a need for review of defensive driving measures.
Loss ratio = $18,000 (losses) ÷ $33,000 (premium) = 55%

_____   _____   _____   _____
Date   Loss Control Director   Date   Chairman Safety Committee

Fig. 15

SUMMARY OF
PROPERTY LOSSES (REAL AND PORTABLE)

Colorscope Incorporated, Permafrost, Arizona
January 1, 1973 through July 1, 1973

COLORSCOPE INCORPORATED
PERMAFROST, ARIZONA
SUMMARY OF PROPERTY LOSSES

1973
Report Period
1-1-73 to 7-1-73

DEPARTMENT	REAL								PORTABLE						Total No. of Losses
	Fire	Wind	Lightning	Flood	Hailstone	Earthquake	Vandalism	Rainstorm	Products	Audio-Visual	Office Machines	Special Tools	Instruments	Money and Securities	
Administration							1					1			2
Advertising				1							1				2
Art															0
Chemical							1								1
Construction					1										1
Engineering															0
Maintenance															0
Manufacturing															0
Quality Control													1		1
Receiving								1							1
Sales															0
Security															0
Service															0
Shipping										1					1
Warehouse								1							1
Totals by Type	0	0	1	1	0	0	4	0	0	1	1	0	1	0	10

Total No. of Occurrences	10
Total Dollar Loss	$13,850
Recovered Amounts	5,000
Actual Dollar Loss	8,850
Insurance Premium	23,700
Loss Ratio to Date	37%

Remarks:
1. Rainstorm losses involved roof leaks during February rains over one-week period. Total damage: $6,300.
2. Products loss involved shipment of coating material in tanker of contact hauler $5,000 loss. Recovered by shipper's insurance payment.
3. Money loss $300 and special tool loss $600 were the result of a burglary on March 17, 1973.
4. Audio-visual equipment loss $750. Items taken from salesman's automobile.
5. Flood damage of $900 when material for outdoor sign was involved in washout on leased property.

_____   _____   _____   _____
Date   Loss Control Director   Date   Chairman Safety Committee

Fig. 16

CHAPTER SIX

The Standard Procedure Instructions

It is difficult to determine the point in time in our nation's history when the need for Standard Procedure Instructions in industry was recognized. We do know that in 1910 the National Association of Manufacturers made efforts to hold the employer responsible for a share of the economic loss suffered by an employee because of a work injury. Much earlier, however, Benjamin Franklin, as an insurance man, had expressed concern over the act of carrying live burning embers from one stove to another in the next room. The loss of many mercantile and dwelling structures by fire in those early days was, in fact, due to noncompliance with accepted practices.

At the start of the industrial revolution in this country, early in the nineteenth century, some disastrous losses related to fires and industrial injuries occurred. These losses had some financial burden attached that was not, for one reason or another, wholly recoverable through insurance. There were many companies that could not sustain their losses and went out of business. Companies large enough to regain a semblance of balance did, in fact, seek methods to reduce the probability of recurrence. Their initial action was independent except for the direct assistance that was available from the insuror. In this connection, there was very little in the way of positive good that spilled over to the other companies in other cities or states. Each company developed its own corrective measures as a result of its own loss experience.

The National Association of Manufacturers, insurance companies, and safety organizations have accomplished much in identifying the fundamental problem of losses in the home and industry. They have, for years, offered the loss control measures that can be applied by almost any commercial, educational, industrial, mining, or manufacturing enterprise. The cost for the materials and service has always been nominal and within the budget range of even the small company. Those companies taking advantage of the services offered by these organizations have improved—but have not fully resolved— their loss problems. The addition of special staff to improve the company resources concerned with facility, processes, and people has provided further loss reduction with improved standards. However, until OSHA, the safety program was generally an "on again, off again" activity.

Since the early twentieth century, we have believed that 88 percent of industrial accidents resulted from the "faults of employees" in the form of "unsafe acts," and even a large portion of the remaining 12 percent were related to job conditions attributed to poor work practices. This belief has continued over the years and has contributed much to the present day safety rules, standard operating practices, safety instructions, etc. However, until the Occupational Safety and Health Act of 1970, there was no total national acceptance of specific standards to reduce the probable loss of life and property.

The compliance standards published in the *Federal Register* are "the law." They do, in fact, impose upon the employer the responsibility for job conditions and work practices. The federal standards are readily understood by the safety professional and are used as the basis for formulating Standard Procedure Instructions that will comply fully with the responsibility placed upon the employer by the federal government.

The importance given to job conditions in the *Federal Register* is echoed by safety representatives throughout the land. However, we again emphasize that the employer who is in compliance with the Occupational Safety and Health Act of 1970 is in the best position ever to eliminate the employee (or employees) who insists on engaging in substandard work practices. We hasten to point out that a Standard Procedure Instruction must be in force and the employer must be prepared to prove that instruction was, in fact, provided for compliance.

The Standard Procedure Instructions for each company and each location should be structured to the particular need for any function. They should provide a specific action for the controlled effort of the entire workforce. They should collectively contribute to the primary objectives of the company for development of people, growth, and profit yield.

STANDARD PROCEDURE INSTRUCTION: NO. 1

Subject: Emergency Reporting					Date June 1, 1974

GENERAL

The early response to an emergency by trained specialists is of paramount importance to the success of a life safety and property protection program. The specific treatment areas are indexed here for supervisory reference. Specific emergency call numbers should be widely distributed in all areas of the activity.

EMERGENCY CALL LIST					PHONE NO.

1. *Security*
   Security department
   City police department
   County sheriff's department
   State department of public safety

2. *Fire*
   City fire department

3. *Ambulance*
   Medical office or dispensary

4. *Maintenance*
   All crafts on call
   Off-site emergency maintenance
   At night, holidays, or on weekends,
   call the appropriate security office

STANDARD PROCEDURE INSTRUCTION: NO. 2

Subject: Accident Reporting   Date June 1, 1974

Losses which must be reported shall include all injuries to personnel and damaged property, facilities, equipment, vehicles, and cargo.

A. FIRST-AID INJURIES

All injuries of a minor nature shall be reported to the supervisor, recorded in the first-aid report log, treated with a mild antiseptic, and adequately protected against further injury or infection. First-aid WILL NOT be administered to the eyes for any traumatic injury, nor will aid be given by internal processes. Injuries requiring such treatment shall be referred to a competent medical doctor.

B. SERIOUS OR DISABLING INJURIES

All serious injuries shall be given prompt medical attention, and every effort shall be made to protect the injured from further injury and/or exposure until medical authority is in charge.

A Supervisor's Investigation Report will be initiated by the supervisor most closely associated with the activity. In addition, the necessary accident report form will be completed as required by the state industrial commission and forwarded to the insurance claims office designated for each location. A copy of this report will be forwarded to the loss control department without delay. The record-keeping, as required by OSHA, will be maintained at the loss control department.

C. PROPERTY DAMAGE

All damage to facility property (fire, windstorm, lightning, flood, vandalism, etc.) will be reported to the security division of the loss control department and the necessary local authority. A Supervisor's Investigation Report will be initiated. A copy of each report will be sent to the loss control department.

D. MOTOR VEHICLE ACCIDENTS

All motor vehicle accidents will be reported to the security division of the loss control department designated by each activity and as required by the state and local authority. The supervisor most closely associated with the activity or the loss control department will prepare a Supervisor's Investigation Report. A copy of each report will be sent to the loss control department.

E. CARGO LOSSES

All cargo losses resulting from theft or damage will be reported to the security division of the loss control department, and the supervisor most closely associated with the activity will prepare a Supervisor's Investigation Report. A copy of each report will be sent to the loss control department.

STANDARD PROCEDURE INSTRUCTION: NO. 3

Subject: General Housekeeping Date June 1, 1974

GENERAL

1. Good housekeeping is the responsibility of all personnel involved in company activity and is essential for accident prevention. The overall efficiency of any department is directly related to its state of orderliness and cleanliness. Poor housekeeping is not only a direct contributor to many injuries and health hazards, but is a primary cause of fires. In every environment, good housekeeping is essential in any accident prevention program.
2. It is important to plan for equipment arrangement, material storage, and environmental conditions to be encountered in any structure. These items are essential elements of good housekeeping.
3. Suitable metal waste containers shall be used to collect industrial scrap and office waste materials. All containers shall be emptied at least once each day. The disposal of cigars, cigarettes, and pipe ashes in this type of container *is prohibited*. Containers used to collect combustible waste, i.e., oily rags, paint-soaked rags, and similar flammable materials, shall be cans with self-closing metal lids, and will be emptied on a regular daily basis.
4. Excelsior, straw, shredded paper, and other packing materials shall be stored in an isolated fire-resistant area. If these packing materials are to be stored in shops for immediate use, they will be stored and maintained in metal or metal-lined, covered containers. Under no circumstances will the quantity of material stored in the containers be such that the covers cannot be properly closed.
5. Floors in the building shall be maintained so they are smooth, clean, and free of obstructions and slippery materials. Under no circumstances will flammable liquids be used in the cleansing process. When toxic cleansing agents are used, adequate ventilation will be provided to remove vapors. Extreme caution shall be exercised to avoid excessive waxing or polishing of floors. Slippery floors contribute to falls.
6. All windows shall be kept clean and maintained in good working condition at all times. Broken glass shall be immediately reported for repair.
7. Corridors and aisles shall be clearly defined and kept free of any hazardous obstructions. Corridors and aisles shall be maintained at widths established in the National Fire Prevention Life Safety Code number 101. Aisle spaces in storage areas shall be kept clear and free from obstructions for easy access to fire-fighting equipment and to enable firemen to reach the fire. Areas shall be kept clear around sprinkler heads and control valves, fuse boxes, and electrical switch panels. These areas shall be clearly identified.
8. Poorly lighted areas are breeding grounds for poor housekeeping and

increase the accident potential. All machinery, work benches, aisles, stairways, and rooms will be adequately and properly lighted.
9. Vending machines can contribute to overall poor housekeeping. All persons shall be instructed to return bottles to racks and to dispose of papers, cartons, and cups in trash cans. Electrically operated vending machines shall be grounded. They shall not be installed in areas where the danger of explosive gases or vapors is present, unless such equipment is designed for such purposes and complies with the existing National Electrical Codes.

STANDARD PROCEDURE INSTRUCTION: NO. 4

Subject: Safety Apparel and Appliances   Date June 1, 1974

A. EYE PROTECTION

1. All employees and visitors engaged in occupations or present where eye hazards due to flying particles, hazardous substances, or injurious light rays are inherent in the work or environment, shall be safeguarded by means of eye protection in accordance with the Occupational Safety and Health Act of 1970.
2. Safety glasses will be worn at any time any employee is engaged in, or working in the proximity of, operations that give rise to flying particles, dust, grit, metal, liquids, chemicals, etc. Grinding, chopping, chiseling, sawing, chipping, and similar operations are included.
3. Supervisors or department heads may define certain areas within their own departments where eye protection must be worn at all times due to the nature of the work.
4. The design, construction, testing, and use of devices for eye and face protection shall be in accordance with the American National Standard for Occupational and Educational Eye and Face Protection, Z87.1—1968.

B. HEAD PROTECTION

1. Staff and visitors, working or present in locations where the hazards of flying or falling objects or substances are inherent in the work or the environment, shall be safeguarded by means of approved head protection.
2. Helmets that are to be used for protection against impact and penetration from falling and flying objects and from limited electrical shock and burn shall meet the requirements and specifications established in American National Standard Safety Requirements for Industrial Head Protection, Z89.1—1969.

C. PROTECTIVE CLOTHING

1. Staff or visitors working or present in areas that expose parts of their bodies, not otherwise protected, to hazardous or flying substances or objects may be required to use additional specialized

body protection equipment, such as aprons, as required by the Occupational Safety and Health Act.
2. Hand protection may be required for staff or visitors whose hands are regularly exposed to hazardous substances and the possibility of cuts or burns.
3. Foot protection may be required for staff or visitors who are exposed to foot injuries due to hot, corrosive, or poisonous substances, or from falling objects. Safety footwear that is worn must meet the requirements established by the American National Standard for Men's Safety-Toe Footwear, Z41.1—1967.

D. HEARING PROTECTION
1. Staff or visitors exposed to levels of noise which may cause hearing impairment shall be provided with the proper hearing protective equipment. This will usually be an approved type of ear plug, muff, or canal cap, designed to obstruct sound energy from entering the ear canals.

STANDARD PROCEDURE INSTRUCTION: NO. 5

Subject: Maintenance Activities   Date June 1, 1974

A. POWER MACHINERY AND EQUIPMENT
1. Supervisors of maintenance areas shall be responsible for ensuring that only qualified, experienced personnel operate power machinery. Power saws, sharpeners, and similar equipment must have the proper type of safeguards in place when the equipment is in operation. Protective eye equipment *will* be used when personnel are operating machinery which could cause particles to be discharged in such a manner as to cause injury. All machinery shall be properly grounded. Control switches shall be located at the point of operation best suited for equipment control in accordance with the Occupational Safety and Health Act of 1970 and the National Electrical Code.

B. AUTOMOTIVE MAINTENANCE AREA
1. Grease on floors and greasy tools account for a great number of accidents. The practices outlined here shall be considered minimal requirements in automotive shops and repair areas.
   a. Satisfactory housekeeping conditions must prevail at all times.
   b. Lubrication racks and lifts must be kept clean and free of grease and debris at all times.
   c. Drop lights should be equipped with vaporproof globes and shields. The lights must be of an approved type. Under no condition shall the cord be spliced.
   d. Oily and grease-soiled rags shall be kept in self-closing metal containers and removed from the area on a daily basis. Under no condition shall the containers be left uncovered.

      e. Gasoline or other flammable solvents will not be used as a cleansing agent for vehicles, vehicle parts, floors, or for other items associated with automotive maintenance shops.
      f. Smoking shall be restricted to specifically designated areas in the maintenance shops. The smoking area shall be conspicuously marked and designated, and butt receptacles shall be provided. All other areas shall be plainly posted for NO SMOKING, and the ruling will be rigidly enforced.
      g. Vehicle engines or machinery with toxic exhaust vapors will not be operated in enclosed areas without a safe exhaust system being utilized.
      h. Air compressors shall be properly grounded, control switches in proper repair, and stop switches painted red. Drive belts shall be equipped with a guard to prevent an individual, tool, or piece of clothing from being caught in the drive unit.

C. EQUIPMENT ROOMS (ELECTRICAL AND MECHANICAL)
  1. Fires and accidents in equipment rooms are often the result of spontaneous ignition. To prevent losses of this type, equipment rooms, mechanical and electrical, shall be kept clear at all times and will not be used as storage areas for any materials. Oily and grease-soiled rags shall be immediately removed and stored in metal self-closing containers in a location away from equipment rooms.

D. WELDING OPERATIONS
  1. Welding operations performed on our premises must conform to the standards as directed by the Occupational Safety and Health Act of 1970, subpart Q: welding, cutting, and brazing, paragraphs 1910.251–1910.254.
  2. All personnel using high-pressure compressed gas cylinders are encouraged to use the highest degree of sound safe practice in the handling of these cylinders, i.e.,
      a. Empty cylinders must be plainly marked "MT" and removed to a segregated storage area.
      b. Cylinders shall be secured in position, regardless of status (in use or storage), by restraining straps or chains mounted to a rigid surface to prevent tipping.
      c. When cylinders are not in use, make sure that the valve is in the off position and not leaking, and that the head caps are firmly in place.
      d. Only the regulator and gauges designed for the particular gas will be used. Never force ill-fitting gauges and regulators into positions because leaks may appear, creating additional fire hazards.
      e. Always store cylinders with valve end *up* and bonnet installed.

f. When movement of the cylinders is required, transport cylinders with a wheeled cart.
    Do not slide or roll.
g. Store high-pressure gas cylinders in a well-ventilated area and away from excessive heat or ignition sources.
h. Never refill MT cylinders.
i. ICC rules require that hydrostatic testing procedures be used as preventive maintenance to detect cylinder fatigue. Refuse delivery of cylinders having hydrostatic test dates more than five years prior to date of delivery.

STANDARD PROCEDURE INSTRUCTION: NO. 6
Subject: Chemical Laboratories   DATE June 1, 1974

I. RESPONSIBILITY

A. The research director has the primary responsibility for initiating action on these rules in research laboratories. He or she bears primary responsibility for seeing that equipment designed for use in the laboratory is made safe before it is operated. The safety committee will assist the director in fulfilling this responsibility if requested to do so.
B. The senior faculty member in charge of a teaching laboratory bears primary responsibility for the enforcement of safety rules.

II. HOUSEKEEPING

A. The basic responsibility for the housekeeping of each laboratory rests with the person in charge of the area. It is his job to see that persons working in his laboratory area keep the working space in good order, and that unsafe conditions of the working space and utilities are reported and corrected.
B. It is the specific duty of the research director to see that excessive amounts of chemicals and equipment do not accumulate in the laboratories under his direction.
C. The custodian of supplies is responsible for the housekeeping in the stockroom and storage sheds.
D. Some general and specific housekeeping rules are:
    1. Keep benches, dry boxes, and hoods neat. Do not use them for inactive storage.
    2. Bench drawers are to be used for storing equipment and chemicals that you are not using. Return such equipment and chemicals to the stockroom.
    3. Chemical products should not be allowed to accumulate to excess. Periodic housecleaning should result in the disposal of unnecessary and hazardous chemicals. The safety committee will give assistance if requested to do so.

III. Hazardous Operations
   A. *All new operations* and equipment or significant changes in operations or equipment should be approved by the research director.
   B. *Hazardous operations* must be appropriately labeled and guarded to prevent injury to others. Nearby workers should be warned before a hazardous operation is started.
      1. Appropriate safety shielding or barricading will be used when, in the opinion of the responsible person involved, this is required.
   C. Unattended Operations
      1. It is essential that all operations be designed to be "fail safe." Remember that every utility—cooling water, electricity, compressed air, natural gas, ventilation, and tank nitrogen—is subject to failure without warning. Plan for such emergencies.
      2. In the event that a potentially hazardous operation is to continue unattended for a period of hours or overnight, previous approval should be obtained from the research director. Before granting approval, he must satisfy himself that the operation is of a nonhazardous nature and that there is no probability of a dangerous condition occurring as a result of the operation. After approval is obtained, the following precautions should be observed:
         a. An overnight operation card must be posted on the outside of closed doors of the laboratory.
         b. A light must be left on in the room where the reaction or operation is in progress.
         c. Arrangements must be made with the security guards to have the guard look at the area for signs of trouble each time he checks the laboratory building.
IV. Machinery and Power Equipment Hazards
   A. Be thoroughly familiar with machinery and power equipment before attempting to operate it.
   B. Power equipment must not be operated until moving parts such as shaft couplings, gear trains, belt drives, etc., have been guarded.
   C. Some point-of-operation hazards such as drill bits, roll nip points, saw teeth, grinding wheels, and shear points *are not always* subject to guarding. Do not become inattentive while using machines having such hazards.
   D. Gloves, ties, jewelry, and loose clothing can be caught by both smooth and rough rotating machine parts. Do not expose yourself to injury by oversight of such hazards.

V. Procedure for Reporting Injuries, Accidents, Hazards, or Hazardous Practices
   A. A person receiving any injury (no matter how trivial) shall report it without delay to his research director, if possible, or see that the director is advised, and shall then proceed to the infirmary for treatment. The person notified of the injury shall be responsible for seeing that a report is filled out by the injured individual and shall set any accident preventive measures in motion at once.
   B. The procedure for reporting an injury is as follows:
      1. The person having an accident which results in injury fills out an accident report form which he obtains from the school secretary. He gives the completed form to his research director *on the day of the accident,* unless prevented from doing so by the seriousness of his injury. In this latter case his research director will prepare the form for the injured person.
      2. The research director adds his comments, signs the form, and forwards it to the safety committee.
   C. Accidents (noninjury), hazards, and hazardous practices must also be promptly reported by those who observe them to the research director or safety committee. In the case of hazards and hazardous practices which involve the facilities or housekeeping of any particular laboratory, the person responsible for the area should be notified.
      1. It may appear that some accidents, hazards, and hazardous practices are of such a trivial nature that no report is necessary. The point to be remembered is: Will a report of the accident or hazard be of service in detecting and eliminating an unsafe condition? If an unsafe condition exists, a report should be made.
   D. In the case of serious accidents and accidents of obscure origin, none of the evidence should be released until an investigation is made by the proper authorities. Naturally, all steps necessary for protection of life and property take precedence over preservation of evidence.

VI. Flame Hazard
   A. Persons assigned to a laboratory may use an open flame in their own laboratory, providing the following rules are scrupulously observed:
      1. Use the open flame only when necessary and only for the period of time that it is actually needed.
      2. Prior to lighting the flame, remove all flammable liquids from the immediate area where the flame is to be used.

Check all containers of flammable liquids in the area to be sure that they are sealed.
3. Clear the use of the flame in advance with other occupants of the room.

VII. WORKING AFTER REGULAR HOURS
   A. Under normal working conditions at least two persons are required on each floor. Arrangements shall be made between individuals for cross-checking periodically. It is permissible for one person to work alone when doing 100 percent desk work.
   B. Under working conditions other than normal, special rules may have to be formulated. For example, two persons may be required in *one room* if an unusually hazardous operation is in progress. The research director in charge has responsibility for determining whether or not the normal rules are adequate and for specifying additional coverage if, in his opinion, they are not adequate.

VIII. EATING AND SMOKING IN WORK AREAS
   A. Smoking is strictly prohibited except in specifically approved areas.
   B. "No Smoking" signs will be displayed on the doors of all laboratories where smoking is not allowed. "Smoking at Desk Only" signs will be displayed on the doors of laboratories which have been approved for smoking.
   C. Permission to eat at the desk in a laboratory can be obtained from the appropriate research director. Before granting this permission, he must assure himself that the toxicity hazard of the laboratory is sufficiently low to warrant granting this permission.
   D. No food or beverage is to be stored in the laboratories. Store food only in facilities approved for this purpose. *Do not* use facilities designated for chemical storage.

IX. PERSONAL PROTECTIVE EQUIPMENT
   A. CLOTHING
      1. The school is not responsible for damaged clothes, shoes, or jewelry.
   B. EYE PROTECTION
      1. The use of eye protection is required *for all work* other than desk work. This includes both mechanical and chemical work, and whenever one is in proximity to such activity. Choose a protective device in keeping with the harmful potential of the exposure. Goggles and face shields should be available in those laboratories where their frequent use is to be expected.

C. RESPIRATORY PROTECTION
1. Respiratory protection should be worn when one is in contact with harmful dusts or gases. Each person should know the location and operation of this protective equipment.
2. All-service-type canister gas masks are available at several locations in the laboratory halls. The canister masks are suitable for contamination levels of up to 2 volume percent *but do not provide protection if the oxygen level of the breathing atmosphere is below 16 percent.*
3. Oxygen masks are available for protection in atmospheres lacking in oxygen or heavily contaminated by an emergency.
4. *Know the respiratory exposure potential.*
5. *Select an adequate protective device.*
6. *Retreat from the exposure if the device selected fails to protect.*
7. *Don't take a chance. Be sure.*

X. VACUUM EQUIPMENT
A. Any glass equipment operated under vacuum can collapse violently, causing a shower of flying glass. For this reason, appropriate shielding must be used during these operations. For suction filtrations, always use the thick-walled flat-bottom flasks specifically designed to withstand vacuum. Even these are dangerous if cracked or otherwise weakened. Ordinary Erlenmeyer flasks larger than the 50-milliliter size should never be subjected to vacuum.
B. Dewar flasks must be shielded by or wrapped in electrical or cloth tape. Vacuum desiccators must be shielded by complete enclosure in a suitable container.
C. Cold traps of sufficient size and temperature to catch all condensable vapors should be inserted between the system and the vacuum pump. A pressure-measuring device should be installed before the cold trap. Check these cold traps frequently to guard against their becoming plugged by freezing of the material collected in them.

XI. ELECTRICAL EQUIPMENT
A. GENERAL
1. While all research personnel may be familiar in varying degrees with utilities and services, electrical repairs may be made only by those authorized. 110-volt service lines can be a source of serious accidents. Under certain conditions, even contact with 25 volts has caused fatalities.
B. REQUIRED IN ALL LABORATORIES
1. All units must be properly grounded.

2. Do not use defective or damaged wires or equipment.
3. Explosion-proof equipment must be used in those areas where the possibility of flammable atmospheres exists.
4. Open Variacs and other non-explosion-proof electrical equipment should not be used in hoods containing flammable materials.

C. PRECAUTIONS
1. Damp concrete provides a good electrical ground. Never stand on wet floors when operating switches, Variacs, stirrers, etc. Rubber floor mats are recommended where such hazards exist.

D. ELECTRIC SHOCK
1. Report all cases of electric shock, however minor, to your research director.
2. When a fellow worker is unable to release himself from the current source, do not touch him, but pull the switch to stop the current. Or, depending on the circumstances, a belt, rope, or other nonconductor may be used to remove the person or to move the wire or other current source.
    a. If the person is unconscious of has stopped breathing, apply artificial respiration, and keep him as warm as possible.
    b. Contact the infirmary for assistance.

XII. HANDLING GLASSWARE
A. A major portion of laboratory injuries results from wounds inflicted through improper usage of ordinary glassware. Certain basic steps may be followed to eliminate the majority of these injuries. One of the most important steps is to always protect the hands by using either gloves or heavy cloth whenever manipulating glass tubing.
1. Inserting glass tubing into rubber equipment
    a. Be sure that the ends of the tubing are fire-polished.
    b. Never try to force glass tubing into an orifice that is too small.
    c. Lubricate the tubing before insertion. Glycerin is a handy lubricant, although water or stopcock grease may also be used.
    d. Never use excessive force on glass tubing.
2. Removing rubber from glass connections
    a. Do not use excessive force to remove glassware.
    b. Wet the glass tubing with water, and also force some water between the glass and rubber surface. *If lubrication does not loosen the connection, cut the rubber away.*

3. Removing frozen stopcock plugs and stoppers
   a. Avoid the use of force.
   b. Gentle tapping of a frozen stopper with another glass stopper or wooden spatula handle will often break the frozen stopper free.
   c. Immersion of a frozen stopper into warm water may also help free the stopper.
   d. Special assistance may be obtained from the glassblower.
4. Cutting glass tubing
   a. Use a sharp file or cutter. Always hold the tubing near the scratch, and break away from the person. It may be necessary to wet the glass before scratching to obtain a clean break.
   b. Always fire-polish the freshly broken ends.
5. Broken glassware and glass waste
   a. Provide a crock container in the laboratory marked "Glass."
   b. Empty all waste glass of chemicals and dispose of the glass into the special container.
   c. Never place waste glassware in ordinary waste paper containers.

XIII. CORK AND STOPPER BORERS
   A. GENERAL
      1. Cork and stopper borers are knives and can inflict a severe wound if not used properly.
   B. HANDLING
      1. Be sure that the borer blade is sharp, since undue pressure may be necessary if a dull blade is used.
      2. Lubricate the area to be bored with either household oil or glycerin.
      3. Leave the borer clean and ready for use.

XIV. HOOD OPERATIONS
   A. The laboratory hood is capable of containing and disposing of hazardous chemical vapors, fumes, and dust when operated in keeping with its design. In general, hoods will provide inflows of 50 linear feet of air per minute. In several instances higher flow rates are provided to control special hazards. The following operating standards and practices are prescribed to ensure satisfactory operation.
      1. Inflow velocities can be improved by decreasing hood sash opening. It is of particular importance that hood openings be kept at a minimum when a single exhaust fan serves a multiple hood arrangement.

2. Avoid creating strong cross drafts in front of the hood opening. They can pull contaminants into the room. Wind blowing into the laboratory through an open door or window will create cross drafts.
3. Close the sash if a hood is not in use.
4. Arrange equipment as deep in the hood as is conveniently possible. This assists fume capture.
5. Vent equipment toward the back or in an upward direction. Rupture disk and safety relief valve arrangements should include extensions leading through the top of the hood into the intake duct as close to the fan as possible. Secure the free end of the extension.
6. The hood draft may be used to remove hazardous vapors from the laboratory by closing all entrances and opening two hood sashes.
7. The hood fan should be left on around the clock if a hood contains hazardous material that is being vented or might vent. Otherwise, the hood fan should be shut down when the hood is not in use. However, when a single fan serves several hoods, do not shut the fan down before checking with other users.
8. Flammables should not be vented at a rate that will create a flammable mixture in the exhaust system.
9. Poor or questionable hood operations should be reported immediately to the research director.

XV. LABORATORY MANIPULATIONS
   A. There are few compounds that have no adverse effect on the body if sufficient exposure is accumulated. Therefore, keep exposure to all chemicals at a minimum.
   B. Do not use mouth suction to fill pipettes. Apply vacuum from an aspirator bulb or vacuum line.
   C. Do not start siphons by mouth.
   D. Use beaker tongs to transport beakers containing 600 milliters or less of hot liquid.
   E. Avoid handling large glass vessels of any type when they are filled with liquids. If it is necessary to do so, be sure that proper safety precautions are taken to prevent breakage or spillage.
   F. Boiling stones are suggested for laboratory distillations where glass equipment is used.
   G. Volatile solvents when shaken in separatory funnels can develop considerable pressure. Avoid this by frequent venting. This can be accomplished by inverting the funnel and opening the stopcock.
   H. Be cautious in smelling compounds to examine them. Never

hold the nose directly over a container when testing for odor. Before smelling, inhale deeply so that air can be expelled immediately if fumes are irritating. Keep the breathing-in of any compound to an absolute minimum.
- I. Form the habit of washing hands and face often while handling chemicals. Always wash hands before eating or smoking.
- J. Always use small quantities of reactants when a new reaction is attempted for the first time.
- K. Never use an oil bath for heating highly oxidizing substances.
- L. Never carry out a reaction or heat an apparatus in a closed system unless it is designed to withstand pressure. If it is undesirable to open a reaction to the atmosphere, consider use of a bubbler and nitrogen purge to assure no buildup of pressure.
- M. In case of any doubt about the potential violence of a reaction, use a shield or carry the reaction out in a hood with the safety window closed. Arrange the apparatus so that a minimum of body exposure results.

XVI. MISCELLANEOUS SAFETY RULES
- A. Horseplay will not be tolerated in any laboratory.
- B. Escape doors must never be blocked in any way.
- C. There is an upper limit to the size or weight of the loads that can be safely lifted and transported by one person. If in doubt, get assistance.
- D. Talk with the owner before borrowing equipment. Always make certain that you know the condition of the equipment and what possible contamination it might contain. The only way you can be certain of these facts is to talk with the person who last used the equipment before you borrow it.
- E. Each research laboratory should have a card fastened to the outside of the door on which are listed the names of each research student and the telephone numbers at which they can be reached in case of emergency. The name and emergency phone number of the research advisor in charge of the students should also be listed.

STANDARD PROCEDURE INSTRUCTION: NO. 7
Subject: Handling of Chemicals (General) Date June 1, 1974

I. DISPOSAL OF WASTE MATERIALS
- A. The procedure to be followed in disposing of waste chemicals depends upon the type of material involved. The following methods of waste disposal are suggested.
    1. Disposal of small amounts of nonhazardous chemicals is permitted in the laboratory.
        a. Acids and alkalies in small quantities and in dilute solu-

tions may be disposed of in laboratory sinks. Salt solutions and water-soluble organic compounds may be disposed of in the same manner. All water-soluble materials disposed of in sinks should be washed down with plenty of water to prevent clogging of the drains and corrosion of the pipes.

    b. Chemicals which react violently with water should never be dumped into the sinks. First, neutralize these chemicals with the appropriate agents. Then wash them down the sink with water if this is not otherwise prohibited because of other considerations.

    c. Higher-boiling, water-insoluble organic materials (benzene, hexane, etc.) should not be poured in sinks except in very small amounts, and then they should be flushed with water.

    d. Small quantities of relatively nontoxic, low-boiling, water-soluble organic materials may be allowed to evaporate in flame-free, well-ventilated hoods. Volatile materials should never be poured into the sink.

    e. Nonhazardous solids which are not disposed of in sinks may be put into the crock marked "Glass." These crocks are used for the disposal of broken glass, inert waste, solids, and rags.

    f. Paper will be disposed of in wastepaper baskets provided in each laboratory.

2. Special disposal problems: Obviously, there will be disposal problems that cannot be handled by the disposal methods outlined above. These problems must be worked out on an individual basis by the research director.

II. COMPRESSED GAS CYLINDERS

A. GENERAL

1. A compressed gas is defined by the Interstate Commerce Commission as "any material or mixture having in the container either an absolute pressure exceeding 40 pounds per square inch at 70°F, or absolute pressure exceeding 104 psig at 130°F, or both, or any liquid flammable material having a Reid vapor pressure exceeding 40 psig absolute at 100°F."

B. GENERAL RULES

1. Cylinders may be filled only when the owner consents.
2. Each cylinder must bear a proper caution label and should carry a legible identification mark.
3. When empty cylinders are returned, valves should be closed and protective caps installed.
4. Never attempt to repair or alter a cylinder.
5. Never repaint a commercial cylinder. To do so may destroy its identifying paint code.

C. SPECIFIC RULES
1. Cylinders should be stored in a definite assigned location.
2. Cylinders that are not easily carried or controlled should be transported on a special cylinder cart.
3. Cylinders should be transported with cap in place, except in unusual or special circumstances.
4. All cylinders must be secured with a chain or other approved retainer.
5. Compressed gases should be handled only by experienced and properly instructed persons.
6. Always be sure that the proper regulator is used. Never force connections that do not fit.
7. Open cylinder valve slowly. Avoid the use of a wrench on valves equipped with handwheels.
8. Consult reference material for the use of specific cylinders.
9. Keep the quantity of flammable gases in the laboratory at the minimum consistent with *current* requirements. Do not store unused cylinders in the laboratory.

III. HANDLING OF FLAMMABLE LIQUIDS

A. Most organic liquids are flammable in the sense that the vapors will burn or explode under the proper conditions. By definition in the National Fire Codes, a *flammable liquid* is any liquid having a flash point below 200°F and having a vapor pressure not exceeding 40 pounds per square inch absolute. *Flash point* is the minimum temperature in degrees Fahrenheit at which a flammable liquid will give off flammable vapor as determined by a specific test procedure and apparatus.

B. Flammable liquids present serious fire and explosion hazards when stored or handled improperly. The following rules should be observed:
1. Large quantities of flammable liquids (drums, cans, and bottles) should be stored in the solvents storage area.
2. Outside storage cabinets for flammable solvents should be used whenever possible.
3. Nonglass containers (either plastic or metal) should be used whenever possible. Remember that some solvents will dissolve certain plastics.
4. Solvents for which extreme dryness or reagent purity are not required should be stored in safety cans with flash screens.
5. Do not transport flammable liquids in open containers.
6. Keep as little flammable liquid in laboratories as is consistent with normal laboratory work. Only solvents used in active laboratory programs should be stored in the laboratory.
7. The maximum amount of flammable liquid *stored in glass*

in any one laboratory is limited to 5 gallons for laboratories of 250 square feet or less, and to 10 gallons for all larger laboratories. The maximum amount of flammable liquid *stored in metal containers* should not exceed 5 gallons for any one solvent in each laboratory.
   8. For equipment containing flammable liquids, a leakproof catch pan should be provided to confine accidental spills or leakage.
   9. Do not remove volatile organic solvents from a hot plate, except in a hood.
   10. When flammable liquids are being transferred in air, provide adequate ventilation of the area to prevent buildup of an explosive mixture in the air.
   11. Drums or metal vessels from which flammable liquids are being drained must be grounded and bonded to the receiver.
   12. In areas where there is a possibility of flammable vapors, electrical equipment must be explosion-proof.
   13. Vent lines from equipment should be directed outside or into a hood.
IV. DRYING OF SOLVENTS
   A. This operation often is one of the most difficult and hazardous of laboratory operations. The problem is due to both the flammable nature of most solvents and the reactivity of drying agents. If highly effective dehydration is desired, the drying agent must be very reactive with water and often is also pyrophoric. Separation of dehydrated solvent from the reaction mass then may require a distillation step, which can be hazardous if unstable sludges are allowed to form and overheat.
   B. A number of serious explosions and fires have been experienced during purification of solvents. A critical inspection of any drying operation using highly reactive reducing agents or distillation steps should be made in the interest of laboratory safety.
   C. Strong reducing agents are often used to effect maximum dehydration of solvents. Their use must be limited, however, to solvents with which they will not react. Examples of these reactive compounds are lithium aluminum hydride, sodium borohydride, calcium hydride, magnesium hydride, sodium dispersions, etc. Dehydrite, a commonly available drying agent, is magnesium perchlorate. It will react violently with organic materials.
   D. In practice, it is best to avoid adding the drying agent to the solvent in large excess of the amount needed for complete water removal. The mixture is stirred for several hours or allowed to stand until hydrogen evolution ceases. The container must be vented during this period to prevent rupture, or stoppers being blown out. Separation of dried solvent by vacuum distillation

is preferred to atmosphere distillation because of the lower pot temperatures required. Some of the complex metal hydrides used have fairly low decomposition temperatures and can undergo explosive decomposition if they are overheated.

1. Precautions
    a. For solvents to remain anhydrous, containers should be tightly sealed. Ordinary caps breathe over a period of time and allow air and moisture to enter. Storage in a dry box does not necessarily provide full protection against this problem.
    b. Solvents stored over sodium ribbon will not remain anhydrous if allowed to breathe. However, the caps should not be tightly closed during the drying period because pressure may increase dangerously.
    c. Distillation conditions should be kept as mild as possible and should not proceed to the point where the sludge is not perfectly fluid. Concentrated systems approaching an oxidant-reductant balance should be avoided. It should be recognized that ethers and certain other chemicals form peroxides and should therefore never be distilled to dryness.
    d. Sample sizes of highly flammable solvents e.g., ethers, should be kept to a minimum. One-liter batches are a reasonable maximum for laboratory use.
    e. Safety equipment should be used in the drying and distillation operations. Safety shields are recommended for all distillations, and the more hazardous systems should be enclosed in restricted areas, e.g., hoods.

V. PYROPHORIC MATERIALS
    A. Pyrophoric materials are any materials which, upon exposure to air, have the possibility of spontaneously igniting either immediately or after standing for a short period of time. Such materials are usually liquids and include the lower molecular weight aluminum, zinc, and boron alkyls, alkyl hydrides, or alkyl halides. Pryrophoric solids are also known, such as lithium and sodium aluminum hydrides and finely divided dispersions of metals, such as lithium, sodium, zinc, magnesium, aluminum, etc.
        1. Properties
            a. Since properties vary widely, it is the responsibility of the individual concerned to obtain all available background information bearing on the particular material being used. Some materials (such as trimethylaluminum) are pyrophoric immediately under almost all conditions. Others are pyrophoric only under unusual circumstances. However, the amount of surface area exposed, temperature, and

agitation are important factors. A given compound may not ignite immediately when exposed to air, but shaking or agitating the surface may induce quick flammability. The aluminum-containing pyrophoric materials are also extremely active toward water. Many are explosive when they come in contact with water.
  b. The tendency of these materials to ignite is considerably diminished by diluting them with inert solvents. At 15 to 20 percent concentrations in hydrocarbons they usually become nonpyrophoric, but they still react with oxygen or moisture at a moderate rate. The more volatile the solvent, the more dilution is required to prevent ignition. It should be remembered, however, that if ignition does occur, the presence of the solvent will increase the damage. (The hydrocarbons are flammable.)
2. Storage
  a. Large quantities of pyrophoric materials should be stored in metal containers and kept in a cool, dry area away from cylinders of oxygen or other oxidizing agents. In the laboratory, small amounts may be kept in almost any container that can be kept tightly closed. The container should be stored in a nitrogen box to prevent a fire in the event of an accidental opening. When glass containers are used, always keep them in a supplemental metal container. Use a nonflammable stabilizing material such as vermiculite, dry sand, or glass wool between the metal and the glass container. Special care should be taken to prevent ground glass stoppers from becoming frozen. Sanitab stoppers provide satisfactory substitutes in most cases.
3. Handling
  a. The prime factor in manipulation of pyrophoric materials is prevention of all atmospheric contact. This should be kept in mind continually and appropriate procedures must be devised. Most storage cylinders are fitted with a dip tube connected to one of a pair of valves. Applying nitrogen pressure through the other valve will force the material through tubing into a container that must be thoroughly dried and flushed with nitrogen beforehand. Other transfers, involving smaller amounts, are best carried out in the nitrogen box.
  b. In handling small amounts of pyrophoric material for reactions or tests in the laboratory, a syringe can often be used. Inject the compound directly into the apparatus through a rubber serum cap. For this procedure, the small exposure at the tip of the needle is negligible, although for very active materials it is advisable to cap the needle

with a small rubber stopper. For use of larger amounts in the laboratory, dropping funnels or flasks may be charged in the nitrogen box. These are stoppered and carried to the apparatus. With a vigorous flow of nitrogen issuing from the apparatus, the container may be quickly unstoppered and the flask or funnel connected to the setup.
   c. In carrying out any reactions involving these reactive materials, possible accidents should be anticipated. If any appreciable quantity of material is involved, a metal pan or other container should be placed under the apparatus to contain the reactants should breakage occur. Flammable substances should be removed from the vicinity of the experiment in order to avoid spread of the fire from the pan.
   d. For the aluminum and sodium-containing materials, thought should be given to the use of coolants other than water. The presence of water in the immediate vicinity of the experiment should be avoided.
4. First aid
   a. Most pyrophoric materials cause severe burns when they come in contact with the skin. Even dilute solutions will cause burns because of reactions with moisture and oxygen of the skin, even though ignition does not occur. First aid consists of prompt removal of the material by copious flushing with water *(do not use alcohol or acetone)* and removal of contaminated clothing. The burns should then be treated as thermal burns.

STANDARD PROCEDURE INSTRUCTION: No. 8

Subject: Walking and Working Surfaces   Date June 1, 1974

   I. WORKING SURFACE
      A. Floor surfaces in the workplace, passageways, storerooms, and service rooms shall be kept in good repair. They must be free from holes, splinters, and loose boards. All surfaces must be kept clean and orderly.
      B. All aisles and passageways shall be kept clear. There shall be no obstruction across or in aisles that could create a hazard. Aisles and passageways which are considered permanent shall be appropriately marked.
      C. Special-purpose flooring and surfaces shall be used in refrigerated compartments, wet process areas, and wherever drainage is necessary. Mats, gratings, duckboards, and nonskid material are considered to be special-purpose floor covering.
      D. Openings and holes in floors shall be guarded by either a standard railing with toeboard on all exposed sides or a floor hole

cover of standard strength hinged in place. Typical situations are: ladderways, hatchways, floor chutes, manholes. All temporary openings must be attended or protected by standard railings.

E. Open-sided floors, platforms, and runways 4 feet or more above the floor or ground level shall be guarded by a standard railing on all open sides with toeboards to prevent materials from falling.

F. Special hazards are created by open-sided floors which are in close proximity to dangerous equipment such as tanks containing dangerous chemicals. These open-sided floors shall be guarded with standard railings with toeboards or an enclosed screen solid construction.

II. RAILING SPECIFICATIONS

Every flight of stairs having four or more risers shall be equipped with standard railings.

A. Standard railings shall consist of a top rail, intermediate rail, and posts. The vertical height shall be 42 inches from the top of the rail to the floor. The intermediate rail shall be halfway between the top rail and the floor.

B. For wooden railings, the posts shall be of at least 2 $\times$ 4-inch stock spaced not to exceed 6 feet, with top and intermediate rails of at least 2 $\times$ 4-inch stock. Posts may be spaced on 8-foot centers if the top rail is made of two right-angle pieces of 1 $\times$ 4-inch stock—the top rail is to be surfaced.

C. When pipe railing is used, the posts, top, and intermediate railing shall be at least 1½ inches nominal diameter, with posts spaced not more than 8 feet on centers.

D. When structural steel or other metals of equal bending strength is used, the posts, top, and intermediate rails shall be made of 2 $\times$ 2 $\times$ ⅜-inch angles with posts spaced not more than 8 feet on centers.

E. Railings of all types must be capable of withstanding a load of at least 200 pounds applied in any direction at any point on the top rail.

F. Other types, sizes, and arrangements could be acceptable if they meet the height as well as the strength requirements of rails.

III. RAILINGS

A. A stair railing shall be of similar construction to the standard railing, but the vertical height shall not be more than 34 inches nor less than 30 inches from the upper surface of the top rail to the surface of tread in line with the face of the riser at the forward edge of tread.

B. Handrails shall be mounted directly on a wall or partition using brackets on the lower side of the handrail so that the smooth surface along the top and sides is not obstructed. Handrails shall be either rounded or shaped in such a way that they furnish an adequate handhold and the ends do not constitute a projection hazard.
C. The height of handrails shall be not more than 34 inches nor less than 30 inches from the upper surface of the handrail to the surface of tread, in line with the face of the riser or to the surface of the ramps.
D. Handrails of hardwoods shall be at least 2 inches in diameter. Handrails of metal pipe shall be at least 1½ inches in diamter.
E. Brackets shall be long enough to provide at least 3 inches of clearance between the handrail and wall or any projection. The spacing shall not exceed 8 feet.
F. Completed mounted handrails must be capable of withstanding a load of at least 200 pounds applied in any direction at any point on the rail.
G. Every flight of stairs having 4 or more risers shall be equipped with standard stair railings or standard handrails. The width of the stairs, measured clear of all obstructions except handrails, shall determine railing requirements.
H. Stairways less than 44 inches wide with both sides enclosed require at least one handrail; with one open side, at least one stair railing on the open side; with both sides open, one stair railing on each side.
I. Stairways more than 44 inches wide but less than 88 inches wide must have one stair railing on each open side. Stairways more than 88 inches wide shall be equipped similarly with one intermediate stair railing approximately midway of the width.
J. Winding stairs shall be equipped with a handrail offset where the treads are less than 6 inches wide.

IV. TOEBOARDS
A. Standard toeboards shall be 4 inches in vertical height from their top edge to the level of the floor, platform, runway, or ramp, and securely fastened with no more than ¼ inch clearance above floor level. They may be made of any substantial material either solid or with openings not over 1 inch in the greatest dimension.
B. Where material is piled to such a height that a standard toeboard does not provide protection, paneling from the floor to the intermediate rail or to the top rail shall be provided.

V. OPENINGS
A. Floor openings and manhole covers may be of any material that meets the following strength requirements:

1. Manhole, conduit, or trench covers and their supports, when located in plant roadways, shall be designed to carry a rear axle load of at least 20,000 pounds.
2. Covers should be made of solid construction, but where there is no exposure to falling materials, grills or slatted covers with openings not over 1 inch in width may be used. Covers should be of nonslip surfaces and set flush. They should not project more than 1 inch above the floor level, and where they do they shall be chamfered. Hinges, handles, and bolts or other parts shall be flush with the floor or cover surface.

VI. LOADING
   A. Conspicuous posting of live loads shall be required in every building or other structure used for industrial or storage purposes.
   B. It shall be "unlawful" to place, or cause to permit to be placed, on any floor or roof of a building, or other structure, a load greater than that for which such floor or roof is approved by the building official.
   C. Safe floor loads shall not be exceeded. For water-absorbent commodities, normal floor loads should be reduced to take into account the added weight of water which can be absorbed during fire-fighting operations.

VII. LOAD-BEARING SURFACES OTHER THAN FLOORS
   A. Areas of a cab roof shall be capable of supporting without permanent distortion the weight of a 200-pound man where necessary.
   B. Portable and powered dockboards (bridge plates) shall be strong enough to carry the load imposed on them. The carrying capacity should be plainly marked.
   C. Portable dockboards (bridge plates) shall be secured in position, either by being anchored or by being equipped with devices that will prevent them from slipping.
   D. All types of dockboards (bridge plates) should have a high-friction surface, designed to prevent employees or trucks from slipping.
   E. The sides of all dockboards (bridge plates) should be turned up at right angles, or other means provided, to prevent trucks from running over the edge.

VIII. LADDERS
   A. The construction and manufacturing specifications for purchase and furnishing of portable wood and metal ladders and construction of fixed ladders shall coincide with the ANSI standards. (See references at the end of this procedure.)

B. Ladders shall be maintained in good condition at all times, the joint between the steps and side rails shall be tight, all hardware and fittings securely attached, and the movable parts shall operate freely without binding or undue play.
C. Ladders shall be stored in such a manner as to provide ease of access or inspection, and to prevent danger of accident when withdrawing a ladder for use.
D. Wood ladders, when not in use, shall be stored at a location where they will not be exposed to the elements, but where there is good ventilation. They should not be stored near radiators, stoves, steampipes, or other places subject to excessive heat or dampness.
E. Ladders shall be inspected frequently, and those which have developed defects shall be withdrawn from service for repair or destruction and marked "Dangerous—Do Not Use."
F. Portable rung and cleat ladders shall, where possible, be used at such a pitch that the horizontal distance from the top support to the foot of the ladder is one-quarter of the working length of the ladder. The ladder shall be so placed as to prevent slipping, or it shall be lashed, or held in position. Ladders shall not be used in a horizontal position as platforms, runways, or scaffolds.
G. Portable ladders shall be so placed that the side rails have a secure footing. The top rest for portable rung and cleat ladders shall be reasonably rigid and shall have ample strength to support the applied load.
H. Ladders shall not be placed in front of doors opening toward the ladder unless the door is guarded.
I. On two-section extension ladders the minimum overlap for the two sections in use shall be as follows:

Size of ladder, feet	Overlap, feet
Up to and including 36	3
Over 36, up to and including 48	4
Over 48, up to and including 60	5

J. Ladders used to gain access to a roof or other elevated areas shall have the top of the ladder extend at least 3 feet above the point of support.
K. If a metal ladder tips over, inspect the ladder for side rail dents or bends, or excessively dented rungs; check all connections between rungs and side rails; check hardware connections; check rivets for shear and other harmful structural damage.
L. Safety shoes of good substantial design shall be installed on all ladders. Where ladders with no safety shoes or spikes are used on hard, slick surfaces, a foot-ladder board should be used.
M. Users shall be instructed in the necessary safety measures when

metal ladders are used in areas containing electric circuits to prevent short circuits or electric shock.

IX. SCAFFOLDS
- A. Specifications for the construction, manufacture and furnishing, and erection of scaffolds and scaffolding materials shall coincide with the ANSI standards listed on page 99.
- B. Scaffolds shall be furnished and erected for persons engaged in work that cannot be done safely from the ground or from solid construction. Ladders used for such work shall conform to Subpart D of the *Federal Register* "Walking and Working Surfaces," paragraphs 1910.25 and 1910.26.
- C. The footing or anchorage for scaffolds shall be sound, rigid, and capable of carrying the maximum intended load without settling or displacement. Unstable objects such as barrels, boxes, loose brick, or concrete blocks shall not be used to support scaffolds or planks.
- D. Guardrails and toeboards shall be installed on all open sides and ends of platforms which are more than 10 feet above the ground or floor except for scaffolding wholly within the interior of a building and covering the entire floor area of any room therein and not having any side exposed to a hoistway, elevator shaft, stairwell, or other floor opening.
- E. Scaffolds and other devices mentioned or described in this section shall be maintained in safe condition. Scaffolds shall not be altered or moved horizontally while they are in use or occupied. Any scaffold damaged from any cause shall be repaired immediately.
- F. All planking or platforms shall be overlapped (minimum 12 inches) or secured from movement. An access ladder or equivalent safe access shall be provided. Scaffold planks shall extend over their end supports not less than 6 inches nor more than 18 inches.
- G. The poles, legs, or uprights of scaffolds shall be plumb, and securely and rigidly braced to prevent swaying and displacement.
- H. Materials being hoisted onto a scaffold shall have a tag line.
- I. Overhead protection shall be provided for persons on a scaffold exposed to overhead hazards.
- J. Scaffolds under which persons are required to work or pass shall be provided with a screen between the toeboard and guardrail. The screen will extend along the entire opening and consist of No. 18 U.S. Standard wire gauge, ½-inch mesh or the equivalent.
- K. Employees shall not work on scaffolds during storms or high winds.

L. Employees shall not work on scaffolds which are covered with ice or snow unless all ice or snow is removed and the planking is sanded to prevent slipping.
M. Tools, materials, and debris shall not be allowed to accumulate in such quantities as to cause a hazard.
N. All pole scaffolds shall be securely guyed, or tied, to the building or structure. Where the height or length exceeds 25 feet, the scaffold shall be secured at intervals not greater than 25 feet vertically and horizontally.

REFERENCES

ANSI Standard A12.1, "Safety Requirements for Floor and Wall Openings, Railings, and Toe Boards."
ANSI Standard A58.1, "Minimum Design Loads in Buildings and Other Structures."
ANSI Standard A64.1, "Requirements for Fixed Industrial Stairs."
ANSI Standards A14.1, "Safety Code for Portable Wood Ladders."
ANSI Standard A14.2, "Safety Code for Portable Metal Ladders."
ANSI Standard A14.3, "Safety Code for Fixed Ladders."
ANSI Standard A10.8, "Safety Requirements for Scaffolding."

(Source of all references in Standard Procedure Instructions 8 to 18: *Inspection Survey Guide,* Bureau of Labor Standards, Workplace Standards Administration, U.S. Department of Labor, Washington, D.C.)

STANDARD PROCEDURE INSTRUCTION: No. 9
Subject: Means of Egress　　　　　　　　　　　　Date June 1, 1974

I. Exit Facilities
  A. Exits shall be provided for all industrial occupancies having a capacity of one person per 100 square feet of gross floor area.
  B. Means of egress shall be measured in units of exit width of 22 inches. Fractions of a unit shall not be counted, although an additional 12 inches shall be counted as a half unit.
  C. Exit areas must be clear and measured at the narrowest point. Door travel shall not restrict the effective width at any point. Handrails may not project more than 3½ inches in measured width. A stringer may not project more than 1½ inches in measured width.
  D. The capacity of a unit of exit width shall be as follows: Doors at grade level or not more than 21 inches above or below grade level: one unit for 100 persons. Class A or Class B stairs, outside stairs, or smokeproof towers: one unit for 60 persons. Class A ramps: one unit for 100 persons. Class B ramps: one unit for 60 persons. Escalators: one unit for 60 persons. Horizontal exits: one unit for 100 persons. Street floor exits: one unit for each 100 persons, or one-half unit for each two units of stairway, ramp, or escalator from upper floors; this rule applies to lower levels, such as basements.

E. The minimum width of any exit which is a required travel corridor or passageway shall be 44 inches in the clear.
F. Not less than two exits shall be provided for each and every floor or section; this includes basements used for industrial purposes or related to the operation.
G. A single exit may be permitted for rooms or areas that have a total capacity of less than 25 persons, a direct exit to the street or to an open area outside the building at grade level, and a total travel distance from any point of no more than 50 feet.
H. The minimum width of any path to exit access shall in no case be less than 28 inches.

II. EXIT SIGNS
A. Signs designating exits or paths of travel shall be provided.
B. Exit signs shall be illuminated by a reliable light source having a light field with a value of not less than 5 footcandles on the illuminated surface.
C. Every exit sign shall show the word "Exit" in plainly legible letters not less than 6 inches high, with the principal strokes of the letters not less than three-fourths inch wide.
D. A sign reading "Exit," with an arrow indicating the direction, shall be placed in every location where the direction of travel is not apparent.
E. Exits shall be so arranged that it will not be necessary to travel more than 100 feet from any point to reach the nearest exit, or 150 feet in a building completely protected by an automatic sprinkler system.
F. No dead end may be more than 50 feet deep.
G. A common path of travel may be permitted for the first 50 feet from any point.
H. In no case shall access to an exit be through a bathroom or other room which can be locked.

III. EXIT DOORS
A. Each exit door shall be of the swinging type.
B. It shall swing with exit travel, except when serving a story having a population of not more than 50 persons, provided there are no high-hazard contents.
C. A door giving access to a stairway shall swing in the direction of exit travel. The door shall not block stairs or landings to less than 20 inches. When open it must not interfere with the full use of the stairs.
D. An exit door shall be so constructed as to be readily opened from the side from which egress is to be made. If locks are needed, they shall not require the use of a key from the inside of the building.

- E. Any type of fastening device on an exit door shall be provided with a releasing device which is obvious, even in darkness.
- F. A door with a mechanized opening device must either be provided with a manual system to allow exit travel or be kept closed to eliminate egress.
- G. An enclosure door shall be labeled with a sign reading, "Fire Exit—Keep Door Closed."
- H. All exits shall be free of obstructions or impediments at all times.

IV. HIGH-HAZARD EXITS
- A. High-hazard exits require a modification of the general occupancy conditions.
- B. Slide escapes may be used as required exits.
- C. Every area on each floor shall have at least two exits—in different directions.
- D. Each room shall have two ways of escape.
- E. Rest rooms shall be shielded from high-hazard areas with one exit.
- F. Exits shall be located so that it will not be necessary to travel more than 75 feet from any point to reach the nearest exit.
- G. Open industrial structures shall have exit facilities such that they provide at least one means of escape from any point subject to human occupancy.

V. GENERAL-PURPOSE EXITS
- A. Construction of stairs shall be such that the sum of two risers and a tread (exclusive of projections) is not less than 24 or more than 25 inches; maximum riser height: 8 inches; minimum width of tread: 9 inches.
- B. The rated capacity of stairs shall be 45 persons per minute per 22-inch unit width.
- C. Each square foot of stairs must be designed to support 100 pounds.
- D. Nonslip material shall be used on stair treads and landings.

VI. RAMPS
- A. Class A ramps shall have a width of 44 inches or more, a slope of 1 to $1\tfrac{3}{16}$ in 12, and a capacity, in persons per unit of exit width, of 60 in the down direction and 45 in the up direction.
- B. Class B ramps shall have a width of 30 to 44 inches, a slope of $1\tfrac{3}{16}$ to 2 in 12, a maximum height between landings of 12 feet, and a capacity, in persons per unit exit, of 45 persons in either direction.
- C. Ramps shall support a minimum weight of 100 pounds per square foot of line load.
- D. Ramps shall have nonslip surfaces.

E. The slope of a ramp shall not vary, and its landing shall be level and usable for possible change of travel direction. Guardrails may vary.
F. All openings shall be enclosed or protected to prevent the spread of fire or smoke.

REFERENCES
NFPA no. 101, vol. 4, "Code for Safety to Life from Fire in Buildings and Structures."
NFPA no. 80, vol. 4, "Fire Doors and Windows."
ANSI Standard MH9.1, "Safety Requirements for Longshoring on the Docks."

STANDARD PROCEDURE INSTRUCTION: NO. 10
Subject: Powered Platforms, Manlifts, and Vehicle-Mounted Work Platforms   Date June 1, 1974

I. POWERED PLATFORMS
   A. Powered platforms are divided into two basic types for applying this standard: Type F and Type T.
   B. Type F equipment requirements state that working platforms must be suspended by at least four wire ropes and must be designed to keep their position if one rope fails.
   C. Type F equipment requirements permit only one layer of hoisting rope on the winding drum.
   D. Type F platforms may be either roof-powered or self-powered.
   E. Type T equipment requirements state that a working platform is suspended by at least two wire ropes. It must be designed so that the platform will not fall to the ground if one wire rope breaks.
   F. Employees working on Type T platforms are required to wear safety belts which are attached to a life lever attached to the platform or building structure.
   G. Type T platforms may be either roof-powered or self-powered.

II. ROOF CAR
   A. The horizontal movement of a powered traversing roof car shall not exceed 50 feet per minute.
   B. Restriction of movement is required.
   C. Horizontal movement shall be positively controlled.
   D. A positioning device shall be provided to assure proper position both for vertical travel and during storage.
   E. Normal limits of travel shall be determined by mechanical stops. The stops shall be capable of withstanding a force equal to 100 percent of the inertial effect of the roof car in motion with traversing power applied.
   F. Operating devices shall be provided on the roof car. If more than one operating device is used, it should be so arranged that it is possible for only one device to be operated at any one time.

G. In order to operate the car, all protective devices and interlocks must be correctly positioned.
H. The roof car shall be continuously stable; this corresponds to overturning by 125 percent of the rated load, plus maximum dental overtension of the wire ropes suspending the working
I. The roof car and its supports shall be capable of resisting accidental overtension of the wire ropes suspending the working platform, and the calculated value for safe operation shall include the effect of one and one-half times the normal value.
J. The motor shall stall if the load at any time is greater than three times that required to lift the platform with its rated load.
K. Safe access shall be provided.
L. Maintenance areas shall be provided as well as a secured storage position.

III. WORKING PLATFORMS
A. The construction of a working platform shall be of a girder or truss type adequate to support its rated load at any position of loading.
B. Load rating plates shall appear in a conspicuous place with letters and figures having a height of at least one-fourth inch.
C. The working platform shall have a minimum net width of 24 inches with guardrails and 4-inch toeboards.
D. Open spaces shall be such that they will reject a ball 1 inch in diameter and withstand a load of 100 pounds applied horizontally over any area of 12 square inches.
E. Flooring shall be of a nonskid type and if open shall reject a $9/16$-inch-diameter ball.
F. Access gates shall be self-closing and self-locking.
G. The emergency operating device shall be mounted in a locked compartment (break glass receptacle) and shall have a legend mounted thereon reading, "For Emergency Operation Only. Establish Communication With Personnel on Working Platform Before Use."
H. Working platforms shall be grounded.
I. Regulations shall meet the design, construction, and tests in the ANSI Standard for Vehicle-Mounted Elevating and Rotating Work Platforms.

IV. HOISTING EQUIPMENT
A. Hoisting equipment shall be power-driven in both the up and down directions.
B. Guards and protective devices shall be provided.
C. Motors, brakes, drums, ropes, and connections must meet minimum standards.
D. Directional limit devices shall be provided.
E. Metal data tags shall be securely attached to one of the wire rope fastenings.

F. Emergency stop switches shall be provided in or adjacent to each operating device.
G. Communications equipment shall be provided for each powered platform for use in an emergency.
H. Each installation shall undergo a periodic inspection and be tested at least every 12 months.
I. Each installation shall undergo a maintenance inspection and test every 30 days unless the cleaning cycle is less than 30 days.

V. MANLIFTS

A. Floor openings for both the "up" and "down" runs shall be not less than 28 inches or more than 36 inches in width for a 12-inch belt, not less than 34 inches or more than 38 inches for a 14-inch belt, not less than 36 inches or more than 40 inches for a 16-inch belt; they shall extend not less than 24 inches or more than 28 inches from the face of the belt.
B. The clearance between the floor or mounting platform and the lower edge of the conical guard above it shall not be less than 7 feet 6 inches. If this clearance is not possible, no access shall be provided, and the manlift runway shall be enclosed where it passes through any such floor.
C. When the distance between floor landings is 50 feet or more, one or more emergency landings shall be provided so that a landing is available for every 25 feet or less of manlift travel.
D. On the ascending side of the manlift, floor openings shall be provided with a level guard or cone. The minimum cone angle is 45 degrees with the horizontal and the lower edge of this guard shall extend at least 42 inches outward from any handhold on the belt.
E. Floating type safety cones may be used if they are mounted on hinges at least 6 inches below the underside of the floor and so constructed that a limit switch will be activated should a force of two pounds be applied to the cone edge closest to the hinge. Cone depth need not exceed 12 inches.
F. The entrances and exits at the floor landings affording access to the manlifts shall be guarded by a maze or a handrail equipped with self-closing gates.
G. The floor openings at each landing shall be guarded on the sides not used for entrance or exit by a wall, by a railing and toeboard, or by panels of wire mesh of suitable strength.
H. A clear area shall be provided at the bottom area with limitations.
I. Top clearance of at least 11 feet shall be provided above the top terminal landing.
J. A fixed metal ladder accessible from both the "up" and "down" run of the manlift shall be provided for the entire length of travel of the manlift.

K. Manlift rails shall be secured in such a manner as to avoid spreading, vibration, and misalignment.
L. Total runs of manlifts shall be illuminated.
M. The manlift and its mechanism shall be protected from weather.
N. Mechanical requirements shall be specified; e.g., brakes, belts, speed, strength of supports, handrails, handholds, limit stops, and emergency stops.
O. Manlift instruction shall be posted at each landing or stenciled on the belt; e.g., "Face the Belt," and "Use the Handholds to Stop—Pull Rope."
P. At the top floor, the following illuminated sign shall be posted: "Top Floor—Get Off."
Q. At each landing there shall be a conspicuous sign having the following legend: "Authorized Personnel Only."
R. Operating rules concerning proper use of manlifts shall be given.
S. Manlifts shall be inspected at least every 30 days, and a written record shall be kept of findings.
T. Limit switches shall be checked weekly.

REFERENCES
ANSI Standard A120.1, "Requirements for Powered Platforms for Exterior Building Maintenance."
ANSI Standard A12.1, "Requirements for Floor Wall and Wall Openings, Railings and Toeboards."
ANSI Standard C1, "National Electrical Code."
ANSI Standard A92.2, "Vehicle-Mounted Elevating and Rotating Work Platforms."
ANSI Standard A90.1, "Standard for Manlifts."
ANSI Standard B15.1, "Mechanical Power-Transmission Apparatus."
ANSI Standard A14.3, "Safety Code for Fixed Ladders."
NFPA no. 70, vol. 5, "National Electrical Code."

STANDARD PROCEDURE INSTRUCTION: NO. 11
Subject: Occupational Health and Environmental
Control   Date June 1, 1974

I. AIR CONTAMINANTS
   A. Exposures to gases, vapors, fumes, dusts, and mists by inhalation, ingestion, skin absorption, or contact with any material or substance (1) at a concentration above those specified in the "Threshold Limit Values of Airborne Contaminants" of ACGIH, except for the USASI standards listed in Table I of USDL Title 41 CFR 50-204.50(b) and except for the values of mineral dust listed in Table II of USDL Title 41 CFR 50-204.50(b) and (2) at concentrations above those specified in Tables I and II of USDL Title 41 CFR 50-204.50(b), shall be avoided; alternatively, protective equipment shall be provided and used.
   B. To achieve compliance as listed in "A," feasible administrative

or engineering controls—such as work rotation, working time limitations, process or local exhaust ventilation, and/or process isolation—must first be determined and implemented in all cases. In cases where protective equipment alone or protective equipment in addition to other measures is used as the method of protecting the employees, such protection must be approved for each specific application by a competent industrial hygienist or other technically qualified source.

II. TESTING OF DANGEROUS OR POTENTIALLY DANGEROUS ATMOSPHERES
   A. Confined work spaces shall be tested by a competent person before employees are permitted to enter.
   B. If such tests indicate that the atmosphere in the space to be entered contains (1) a concentration of flammable vapor or gas greater than 10 percent of the lower explosive limit, and/or (2) a concentration of toxic contaminants above the threshold limit value and/or (3) less than 16.5 percent oxygen, appropriate control measures shall be instituted. (Control measures may consist of forced or natural ventilation, use of personal protective equipment, administrative controls, or a combination of these and other effective control techniques.)
   C. A person who must enter a confined work space before a safe level of contaminant(s) is achieved shall be provided with personal protective equipment.
   D. Any person stationed outside of a compartment, tank, or space should have all personal protective equipment immediately at hand and available.

III. VENTILATION
   A. All atmospheres where industrial processes generate and/or give off harmful concentrations of gases, fumes, vapor, dusts, and mists creating harmful exposures by inhalation, ingestion, or skin absorption at or above the threshold limit values shall be safeguarded by all feasible engineering controls.
   Where these controls are not possible or practical and personal protective equipment is used, the equipment must be approved for each application by a competent industrial hygienist or other technically qualified source.
   B. Where mechanical ventilating equipment is used to reduce the concentration of contaminant(s) to acceptable levels, periodic testing shall be performed to maintain safe contaminant levels.
   C. Processes to be exhausted by a single system should be located close together.
   D. Where processes generate different dusts, fumes, or vapors that could result in an explosion when mixed, these substances shall be exhausted separately.

E. Systems handling radioactive materials shall be shielded or isolated to prevent exposure.
F. Dust shall not be permitted to accumulate on the floor or ledges.
G. The intake of air shall be located so that, as far as possible, it does not take in exhausted contaminants from adjacent sources.
H. Exhaust hoods should be so designed that air flow is as low as possible.
I. Every workplace for performing dry grinding, dry polishing, or buffing shall be provided with suitable hoods or enclosures that are connected to exhaust systems.
J. The supply air volume plus the entrained air shall not exceed 50 percent of the exhaust volume.
K. When proper air flow is allowed, the hood static pressure shall be measured and recorded.
L. Control velocities shall conform to regulations in all cases where the flow of air passes the breathing or working zone of the operator and into the hoods. The flow should be undisturbed by local environmental conditions such as open windows, wall fans, unit heaters, or moving machinery.
M. Mechanical ventilation shall be provided when welding or cutting is done.
N. Degreasing or other cleaning operations involving chlorinated hydrocarbons shall be so located that no vapors from these operations will reach or be drawn into the atmosphere surrounding any welding operation.
O. All employees working in, or around, open-surface tank operations must be instructed as to the hazards of the job and also in the use of personal protective equipment and first-aid procedures.
P. Comfortable working conditions are as follows:
1. Thermal comfort exists at 70° to 80°F (20° to 25°C).
2. Relative humidity should be 30 to 60 percent.
3. Air movement should be no more than 25 feet per minute.

IV. OCCUPATIONAL NOISE EXPOSURE
A. Protection against effects of noise exposure shall be provided when levels exceed permissible noise exposure.
B. PERMISSIBLE NOISE EXPOSURES

Duration per day, hr	Sound level, dBA, slow response
8	90
6	92
4	95
3	97
2	100
1	105
½	100
½	110
¼ or less	115

C. When employees are subjected to sound levels exceeding those on the list, feasible administrative or engineering controls shall be utilized.
D. If controls fail to reduce noise to a satisfactory level, personal protective equipment shall be provided.

V. RADIATION HAZARDS
A. No facility or activity shall possess, use, or transfer sources of ionizing radiation in such a manner as to cause any individual to receive a dose in excess of the specified limits.
B. Appropriate monitoring equipment shall be supplied to each employee.
C. Caution signs, labels, and signals shall be used—conspicuously posted. Examples are: "Caution—Radiation Area," "Caution—High Radiation Area," "Caution—Radioactive Materials," and "Caution—Airborne Radioactivity Area."
D. A medical program shall be instituted to provide medical services appropriate to the degree of exposure.
E. All personnel shall be informed of the hazards involved and how to guard against them.
F. Store radioactive materials in such places as to minimize exposure.
G. No facility or activity shall dispose of radioactive materials except by transfer to an authorized recipient or in a manner approved by the Atomic Energy Commission.
H. Personnel protection is required for electromagnetic radiation.
I. Warning symbols shall be posted for radio frequency radiation. The following words shall appear: "Warning—Radio-Frequency Radiation Hazard."

VI. COLOR CODING
A. Colors shall be used to designate physical hazards. Machines or areas not considered hazardous have no specified color coding.
B. Red shall be used to designate
  1. Fire protection
  2. Fire exits
  3. Safety containers for combustibles
  4. Emergency stop bars on machines
  5. Electrical switches (not boxes); switch boxes are orange.
C. Orange shall be used to designate dangerous parts of machines or devices.
D. Yellow shall be used to designate caution.
E. Green shall be used to designate "Safety" and first-aid equipment.
F. Blue shall be used to designate caution as to equipment under repair.

G. Purple shall be used to designate radiation hazards.
H. Black, white, or combinations of black and white shall be used to designate traffic and housekeeping patterns.

REFERENCES

American Conference of Governmental Industrial Hygienists, "Threshold Limit Values of Airborne Contaminants for 1970."
ANSI Standard Z33.1, "Installation of Blower and Exhaust Systems for Dust Stock, and Vapor Removal or Conveying."
NFPA no. 70, vol. 5, "National Electrical Code."
ANSI Standards Z9.2 and Z33.1, "Fundamentals Governing the Design and Operation of Local Exhaust Systems."
ANSI Standard Z41.1, "Men's Safety-Toe Footwear."
ANSI Standard Z43.1, "Ventilation Control of Grinding, Polishing, and Buffing Operations."
NFPA no. 33, "Spray Finishing Using Flammable and Combustible Materials."
NFPA no. 34, "Dip Tanks Containing Flammable or Combustible Materials."
ANSI Standard Z49.1, "Safety in Welding and Cutting."
ANSI Standard Z4.1, "Minimum Requirement for Sanitation in Places of Employment."
USDL Title 41 CFR 50-204.21, "Exposure of Individual to Radiation in Restricted Areas."
USDL Title 41 CFR 50-204.23, "Precautionary Procedures and Personnel Monitoring."
USDL Title 29 CFR 1501.82, "Respiratory Protection."
USDL Title 41 CFR 50-204.29, "Waste Disposal."
USDL Title 41 CFR 40-204.32, "Radiation Exposure Records."
ANSI Standard D1.1, "Safety in Pulp and Paper Mills."
ANSI Standard N2.2, "Radiation Symbols."
ANSI Standard 2.3, "Immediate Evacuation—Sequel for Use in Industrial Installations Where Radiation Exposure May Occur."
ANSI Standard N13.2, "Administrative Practices in Radiation Monitoring."
ANSI Standard Z54.2, "Safety Design and Use of Industrial Beta-Ray Sources."
ANSI Standard Z53.1, "Safety Color Code for Marking Physical Hazards and the Identification of Certain Equipment."

STANDARD PROCEDURE INSTRUCTION: NO. 12

Subject: Hazardous Materials   Date June 1, 1974

I. COMPRESSED GASES

A. Supervisors shall visually inspect to determine that the compressed gas cylinders are in a safe condition.
B. Safety relief devices shall be supplied for compressed gas containers.
C. All handling, storing, and utilization of compressed gases shall comply with Compressed Gas Association Standards. Examples are: Storage rooms shall be dry, cool, and well ventilated with enclosures having a fire resistance of at least 1 hour. Cylinders shall be grouped by type of gas and the groups segregated as to compatibility. Charged and empty cylinders shall be stored separately within the area. Cylinders shall not be stored at temperatures above 125°F. Removable caps shall be kept in place, except when cylinders are in use. Cylinders shall be properly supported to prevent them from being knocked over.

Fuel and oxygen cylinders shall be separated (minimum distance: 20 feet). Cylinders shall not be stored near combustible substances. The valves of the cylinders shall be closed, except when they are in active use.

D. Bulk storage shall be above ground with specific regulations as to location, installation, and venting.

E. Acetylene shall not be generated, piped, or utilized at a pressure in excess of 15 pounds per square inch gauge. Liquid acetylene is prohibited. Generators shall be located in outside houses or specially constructed rooms. Openings of the houses shall not be located within 5 feet of any opening in another building. The rooms shall be built of noncombustible materials having a fire-resistance rating of at least 1 hour. A means of ventilation shall be located at the floor and ceiling levels. Explosion venting shall also be provided. Heating, if necessary, shall be by steam, hot water, or other indirect means.

F. Oxygen cylinders shall be separated from fuel gas cylinders or combustible material a minimum distance of 20 feet or by a noncombustible barrier at least 5 feet high and having a fire-resistance rating of at least a half hour.

G. Storage rooms for liquefied or gaseous hydrogen shall have a fire resistance of at least 2 hours. Openings from the room are prohibited and one wall shall be exterior with explosion venting.

II. FLAMMABLE AND COMBUSTIBLE LIQUIDS

A. Flammable liquids shall mean any liquid having a flash point below 140°F and a vapor pressure not exceeding 40 pounds per square inch at 100°F; flammable liquids shall be divided into two classes: I and II. Class I includes those liquids having a flash point below 100°F, while Class II includes those having flash points above and including 100°F and below 140°F.

B. Combustible liquids shall mean any liquid having a flash point at or above 140°F (60°C) and shall be known as Class III liquids.

C. Sources of ignition shall be controlled. Examples are electrical equipment, wiring, static electricity, open flames, heating services, hot work (welding, cutting), and cleaning or safeguarding small tanks and containers.

D. Flammable and combustible liquids shall be stored in tanks or closed containers, approved for the specific purpose by class of liquid, volume, and location.

E. Safety cans shall be used to store and/or handle flammable liquids in quantities of 5 gallons or less.

- F. Drum storage is permissible for more than 5 gallons but less than 60 gallons if the drums are properly bonded and grounded.
- G. Up to 25 gallons can be stored outside a cabinet or inside a storage room.
- H. Tank storage requires normal venting as well as emergency venting (for excessive internal pressure), special fire extinguishing equipment, special location and protection, installation supports, and special valves (internal or external).
- I. Storage tanks are prohibited inside of buildings.
- J. Warehouse storage requires that the building be at least 50 feet from another building and that exterior walls have at least a 2-hour fire-resistance rating. Containers when piled shall be separated by pallets and no closer than 3 feet from the nearest beam, chord, or girder; nor should they be piled too high from fire protection equipment or fire doors. Aisles at least 3 feet wide shall be provided.
- K. At least one portable fire extinguisher having a rating of not less than 20 B units shall be located not less than 10 feet or more than 25 feet from any flammable liquid storage area.
- L. Liquid transfer shall be accomplished by a closed pipe system, from safety cans, by means of a device drawing through the top of a container, or drawing from a container by gravity through an approved self-closing valve. A distance of at least 25 feet is necessary from other operations.
- M. Enclosed buildings shall be ventilated at a rate of not less than 1 cubic foot per minute per square foot of solid floor area.
- N. A means of disposing of leakage and/or spills shall be provided. (Emergency drainage systems are a case in point.)
- O. Provision shall be made to prevent flammable liquids from entering public sewers, drainage systems, or natural waterways; separators, boxes, or other approved means should be used.

III. SPRAY FINISHING USING FLAMMABLE OR COMBUSTIBLE MATERIALS
- A. Spray finishing operations shall be confined to properly constructed spray booths, rooms, or tunnels so constructed as to have a fire resistance of at least 2 hours and provided with automatic sprinkler protection.
- B. Spray booths shall be substantially constructed of steel, concrete, or masonry; their interior surfaces shall be smooth and continuous without edges. The booths shall be so constructed that residue does not pocket and that cleaning can be easily accomplished. Space within a spray booth having a frontal area greater than 9 square feet should be protected with an automatic sprinkler. A clear space of at least 3 feet should be maintained around spray booths.

- C. Electrical equipment located within 20 feet of the spraying operation shall be installed and maintained in accordance with the National Electrical Code.
- D. The spraying area shall be kept free of all combustible residue whether it be in the booth, ducts, or discharging point.
- E. Mechanical ventilation shall be provided to remove flammable vapors and mists. The discharging location shall be controlled so that life or property is not endangered.
- F. The quantity of flammable or combustible liquids kept in the spraying area shall not exceed the minimum required for a single shift.
- G. Original shipping containers shall not be subjected to air pressure for supplying spray nozzles.
- H. There shall be no open flames or spark-producing equipment in any spraying area, or within 20 feet of the operation unless separated from it by a partition.
- I. Spray nozzle containers shall be of the closed type.
- J. All pressure hoses and couplings shall be regularly inspected.
- K. When transferring liquids from one container to another the containers shall be bonded and grounded.
- L. Solvents for cleaning shall be restricted to those having flash points not less than 100° F.
- M. "No Smoking" signs shall be conspicuously posted.

IV. DIP TANKS FOR FLAMMABLE OR COMBUSTIBLE LIQUIDS
- A. Dip tank operations should be located away from other important operations and construction should be such that it has a fire resistance of at least 2 hours and is provided with an automatic sprinkler.
- B. Egress should not be impaired in case of fire.
- C. Ventilation systems are required, and if any one system fails, the system's controls shall automatically stop any dipping conveyer system.
- D. All dip tanks exceeding 150 gallons of liquid or housing a liquid surface area exceeding 4 square feet shall be protected by an automatic fire extinguisher system or automatic tank cover. Appropriate fire extinguishers shall be provided in the immediate vicinity.
- E. Storage of full or empty containers within the process areas should not exceed the requirements for one operating shift.
- F. Tanks and vats shall be guarded with a standard railing and toeboard.
- G. No electrical equipment shall be in the vicinity of dipping operations which would be subject to splashing or dripping of flammable or combustible liquids.

H. "No Smoking" signs shall be conspicuously posted.

V. EXPLOSIVES
A. No person shall store, handle, or transport explosives or blasting agents when such storage, handling, and transportation of explosives or blasting agents constitute an undue hazard to life and property.
B. All explosives shall be kept in magazines.

VI. LIQUEFIED PETROLEUM GASES
A. All liquefied petroleum gases shall be effectively odorized by an approved agent.
B. Each system shall have its container valves, connectors, manifold-valve assemblies, and regulators approved.
C. Fabrication shall be done in compliance with the code.
D. All containers shall be marked.
E. Location of containers and regulating equipment shall be in compliance with standards.

VII. ANHYDROUS AMMONIA
A. Containers shall be located outside of building or in buildings or sections thereof especially provided for this purpose.
B. Nameplates shall be permanently attached.
C. "Caution" signs shall be in place.

REFERENCES

Compressed Gas Association, P-1, "Safe Handling of Compressed Gases."
NFPA no. 70, vol. 5, "National Electrical Code."
NFPA no. 566, vol. 2, "Bulk Oxygen Systems."
NFPA no. 58, vol. 2, "Storage and Handling of Liquefied Petroleum Gases."
NFPA no. 51, vol. 2, "Oxygen-Fuel Gas Systems for Welding and Cutting."
Compressed Gas Association, G-1, "Acetylene."
NFPA no. 220, "Standard Types of Building Construction."
ANSI Standard B31.1–B31.10a, "Code for Pressure Piping of Gas and Air Piping Systems."
Compressed Gas Association, G-8, "Nitrous Oxide."
Compressed Gas Association, G-4, "Oxygen."
NFPA no. 30, vol. 1, "Flammable and Combustible Liquids Code."
NFPA no. 327, vol. 1, "Cleaning Small Tanks."
NFPA no. 51B, vol. 2, "Cutting and Welding Processes."
NFPA no. 33, vol. 1, "Spray Finishing Using Flammable and Combustible Materials."
NFPA no. 91, "Blower and Exhaust Systems for Vapor Removal."
NFPA no. 13, vol. 1, "Installation of Sprinkler Systems."
NFPA no. 86A, "Ovens and Furnaces."
NFPA no. 34, vol. 1, "Dip Tanks."
NFPA no. 495, vol. 3, "Explosives and Blasting Agents' Code."
ANSI Standard Z48.1, "Method of Marking Portable Compressed Gas Containers to Identify the Material Contained."
ANSI Standard K61.1, "Safety Requirements for the Storage and Handling of Anhydrous Ammonia."

STANDARD PROCEDURE INSTRUCTION: NO. 13

Subject: Fire Protection   Date June 1, 1974

I. GENERAL INFORMATION
   A. The fire department or plant fire brigade shall be called for all fires regardless of size.
   B. All equipment shall be in accordance with requirements of the National Fire Protection Association; local requirements must be equal to or above the NFPA requirements.
   C. Class A fires are fires involving ordinary combustible materials such as wood, cloth, paper, and rubber.
   D. Class B fires are fires involving flammable liquids, gases, and greases.
   E. Class C fires are fires which involve energized electrical equipment where the electrical nonconductivity of the extinguishing media is of importance.
   F. Class D fires are fires involving combustible metals, such as magnesium, titanium, zirconium, sodium, and potassium.

II. PORTABLE FIRE EXTINGUISHERS
   A. Portable units are intended for fires of a limited size and are considered supplementary.
   B. Units shall be fully charged and under operable conditions at all times.
   C. Units shall be conspicuously located and readily accessible along normal paths of travel and near exits.
   D. Locations shall be identified and not obstructed.
   E. Fire extinguishers shall be installed with hangers or brackets, mounted in cabinets, or set on shelves unless they are supported by a wheeled vehicle.
   F. Extinguishers having a gross weight not exceeding 40 pounds shall be installed so that the top of the extinguisher is not more than 5 feet above the floor, while those having a gross weight exceeding 40 pounds shall be installed so that the top of the extinguisher is not more than 3½ feet above the floor.
   G. Directions and information labels shall face out.
   H. Extinguishers shall be suitable for use within a temperature range of at least +40° to +120° F. If temperatures are not within this range, suitable types shall be chosen.
   I. Fire extinguishers shall be provided for the protection of both the building structure (if combustible) and the occupancy hazards contained within the facility.
   J. Rooms or areas shall be graded generally as light-hazard, ordinary-hazard, and/or extra-hazard.
   K. Minimal sizes of fire extinguishers are listed as follows:

Basic minimum extinguisher rating for area specified	Maximum travel distances to extinguishers, ft	Areas to be protected per extinguisher		
		Light-hazard occupancy, sq ft	Ordinary-hazard occupancy, sq ft	Extra-hazard occupancy, sq ft
1A	75	3,000	Not permitted, except as specified in *Federal Register,* "Fire Extinguishers," Par. 4120	Not permitted, except as specified in *Federal Register,* "Fire Extinguishers," Par. 4120
2A	75	6,000	3,000	Not permitted, except as specified in *Federal Register,* "Fire Extinguishers," Par. 4120
3A	75	9,000	4,500	3,000
4A	75	11,250	6,000	4,000
6A	75	11,250	9,000	6,000

L. Maximum travel distances shall not exceed the following specifications:

Type of hazard	Basic minimum extinguisher rating	Maximum travel distance to extinguishers, ft
Light	5B	50
Ordinary	10B	50
Extra	20B	50

M. Extinguishers with Class C ratings shall be required where energized electrical equipment may be encountered.

N. Extinguishers shall be inspected monthly and possibly more often. Check for weight, actuation, and signs of tampering, physical damage, corrosion, or other impairment.

O. Spare extinguishers shall be used when the regular extinguishers are being recharged.

P. Recharging date and maintenance check (initialed) tag shall be securely attached to each unit.

Q. If, at any time, an extinguisher shows evidence of corrosion or mechanical injury, it should be subjected to a hydrostatic pressure test or be replaced.

III. AUTOMATIC SPRINKLER SYSTEMS

A. Every high-hazard occupancy shall have automatic sprinkler protection or such protection as may be appropriate to the particular hazard, including explosion venting for any area subject to dust explosion hazard; all protection shall be designed to minimize danger to occupants in case of fire or other emergency before they have time to utilize exits to escape.

B. When changes are required and shutoff is necessary for a few hours, a temporary water supply shall be made available.
C. Complete sprinkler protection is desirable and recommended.
D. All valves to supply sources shall be sealed open in a satisfactory manner.
E. Identification signs shall be provided; these shall contain information about controls, drains, tests, and alarm valves.
F. Approved corrosion-resistant or special coated sprinklers shall be installed according to local and national codes.
G. Protection against freezing is required.
H. Sprinklers shall not be painted.
I. Water-flow alarms shall be provided on all sprinkler installations.
J. Sprinkler location according to recommended placement is imperative.
K. Clearance of at least 36 inches shall be maintained between sprinkler deflectors and top of storage.

IV. STANDPIPES AND HOSES

A. Standpipes shall be located so that they are protected against mechanical and fire damage.
B. Easy access is required and pipes shall not be more than 6 feet from the floor.
C. Each hose outlet provided for the use of building occupants shall be equipped with not more than 75 feet and preferably not more than 50 feet of approved small fire hose attached and ready to use.
D. Size of nozzles for small hose shall not be larger than ½ inch.
E. Inspection shall be made frequently.

V. FIRE ALARMS

A. Manual fire alarm boxes shall be approved for the particular application and shall be used only for fire-protection signaling purposes.
B. Boxes shall be unobstructed, readily accessible, and located in normal path of exit from the area.

VI. OLDER EXTINGUISHING SYSTEMS

A. When dry chemical extinguishing systems are provided they shall meet the design requirements of the NFPA.
B. When carbon dioxide extinguishing systems are provided they shall meet the design requirements of the NFPA.

REFERENCES

ANSI Standard Z112.1, "Installation of Portable Fire Extinguishers."
NFPA no. 10, vol. 8, "Installation of Portable Fire Extinguishers."

ANSI Standard A54.1, "Carbon Dioxide Systems."
NFPA no. 12, vol. 7, "Carbon Dioxide Systems."
NFPA no. 17, vol. 7, "Dry Chemical Systems."
NFPA no. 68, vol. 9, "Explosion Venting Guide."

STANDARD PROCEDURE INSTRUCTION: NO. 14

Subject: Compressed Gas and Compressed Air
Equipment   Date June 1, 1974

I. COMBUSTIBLE GASES
  A. High-pressure cylinders are those cylinders with a marked service pressure of 900 pounds per square inch or greater, while low-pressure cylinders are those with a marked service pressure of less than 900 pounds per square inch; both are seamless.
  B. All valves shall be checked for proper condition annually, but more frequent checking is suggested. If there is any irregularity in cylinders and valves, they shall be removed from service immediately.
  C. The operating pressure of the vessel shall not exceed the safe limits of the vessel.
  D. Multiple cylinders of less than 300 cubic feet per minute can be stored in the same room providing they follow the following general regulations:
    1. Area shall be properly ventilated.
    2. Cylinders shall be 50 feet from the air conditioning or ventilating system.
    3. Each cylinder shall be 50 feet from any other compressed combustible gas.
    4. Each cylinder shall be 25 feet from flames, electrical equipment, or any other power source.
    5. Each cylinder shall be 25 feet from concentrations of people.
    6. Each cylinder shall be 25 feet from any flammable material.
  E. Compressed gas cylinders, portable tanks, and cargo tanks shall have pressure relief devices installed and maintained.
  F. Only qualified personnel shall be allowed to service safety relief devices.
  G. Vessels shall be subjected annually to a hydrostatic test of one and one-half times the safe working pressure of the vessel.

II. COMPRESSED AIR
  A. Compressed air cylinders shall be protected from bumping and falling by being secured to a permanent support such as a wall.
  B. Air lines shall be reduced in pressure as to not exceed 30 pounds per square inch at the nozzle.
  C. Lines are to be inspected regularly.
  D. No valve of any type shall be placed between the air receiver and its safety valve or valves.

E. Portable air lines shall be stored so as not to hang or extend into machine work areas or traffic patterns.
F. Suitable spring hangers, sway bracing, vibration dampers, etc., shall be provided where necessary.
G. A pressure gauge or a valved connection for a pressure gauge shall be located at the outlet of each pressure reducing valve.
H. Color bands, if used, shall be painted or applied on the pipes and shall be installed at frequent intervals on straight pipe runs and close to all valves.

REFERENCES

Compressed Gas Association, C8, "Standards for Requalification of ICC-3HT Cylinders."
Compressed Gas Association, S1.1, "Safety Relief Devices Standards Part 1—Cylinder for Compressed Gases."
ANSI Standard B-19, "Safety Code for Compressed Air Machinery."
ANSI Standard B 31.1.0, "Power Piping."
ANSI Standard B31.1, "Safety Code for Pressure Piping."
ANSI Standard A13.1, "Scheme for the Identification of Piping Systems."

STANDARD PROCEDURE INSTRUCTION: NO. 15

Subject: Material Handling and Storage   Date June 1, 1974

I. EQUIPMENT OPERATORS
   A. Every operator of powered and motorized special equipment shall be initially examined and certified for proficiency and physical ability. All physically qualified prospective operators not previously examined for proficiency and those operators identified as not meeting our performance standards shall be required to successfully complete an operators' training course.

II. EQUIPMENT
   A. Every unit of powered and motorized special equipment manufactured for general use by our employees shall conform to the ANSI Specifications listed at the end of this Procedure. All such equipment for special use in hazardous locations as defined in the NFPA National Electric Code shall meet the NFPA requirements, as minimum criteria.
   B. Cranes and hoists shall not be loaded beyond the rated load, except for test purposes.
   C. Cranes and hoists shall be frequently inspected for control mechanism maladjustments, control mechanism excessive wear, operability of safety devices, deterioration or leakage in all systems, and deformation of hooks and electrical apparatus.
   D. Cranes and hoists shall be periodically inspected for deformed, cracked, or corroded structural members; loose bolts or rivets; cracked or worn drums, worn, cracked, or distorted pins, bearings, shafts, gears rollers, and locking devices; efficiency of

brake and drum systems; power plant compliance (gas, diesel etc.); deformation of hooks; and worn or damaged tires.
E. Inspection records are required for cranes and hoists.
F. All shore-based cranes and derricks shall be clearly marked to indicate all applicable capacity ratings, based on manufacturer's data for which they are certified.
G. Good housekeeping is required in the cabs of all equipment.
H. A carbon dioxide, dry chemical, or equivalent fire extinguisher shall be kept in the cab or vicinity of the crane. Carbon tetrachloride extinguishers shall not be used.
I. Temporary guarding is required for the swing radius of a revolving crane.
J. All platforms are to be of nonskid materials.
K. Signals shall be discernible or audible at all times. If you do not understand them, stop action.
L. When equipment is operating near electric power lines a 10-foot minimum clearance shall be maintained, except when lines are de-energized.
M. Every industrial truck or tractor, except motorized hand trucks, shall be equipped with a warning horn, whistle, or gong, or other device that can be heard clearly above the normal industrial noises in the working environment.
N. Trucks shall be equipped with driver's overhead guards.
O. Refueling of industrial trucks (except with LP gas) shall be only at locations specifically designated for that purpose.
P. LP-gas-powered trucks shall not be refueled or stored near underground entrances, elevator shafts, or any place where LP gas could collect in a pocket and cause a potentially dangerous condition. Such vehicles shall be refueled outside with engine stopped. Smoking or open lights shall not be permitted in the vicinity.
Q. Battery charging shall be located in an area designated for that purpose with adequate ventilation and a means of flushing and neutralizing spilled electrolyte. A carboy tilter or siphon shall be provided for handling electrolyte. Smoking shall be prohibited in the charging area.
R. Industrial trucks shall not be used in hazardous locations.
S. Fire extinguishers for Class B or C fires shall be kept in close proximity for immediate uses.
T. No one is permitted to ride on a hand truck, and when the truck is not in use the handle must be held in a vertical position.
U. The brake of a motorized hand truck shall be applied and current to the drive motor shall be cut off whenever the steering tongue is in an approximately vertical position.
V. All surfaces for vehicular traffic shall be in first-class condition and free of obstructions; width of aisles for one-way traffic

shall be the width of the widest vehicle or load plus 3 feet, while two-way traffic requires double the widest vehicle or load plus 3 feet.
W. Lanes for aisles and passageways shall be painted on the floor; black, white, or combinations of these two are considered to be housekeeping markings.
X. Guards shall be provided on all types of conveyors with overhead protection where needed. All crossovers, aisles, and passages shall be indicated by suitable signs in conspicuous places.
Y. Emergency shutoff controls are required at designated positions and interlocks where needed.

III. APPARATUS
A. All material handling apparatus and equipment shall be inspected by a qualified person before each use and also at intervals of use, if possible.
B. The eyes of slings shall be so formed or spliced as to maintain the safe working load of the sling throughout.
C. The safe working loads of manila rope and rope slings are determined by the size of the rope and angle of the sling. If a synthetic fiber rope is used to replace a manila rope with a circumference of less than 3 inches, the substitute shall be of equal size.
D. Wire rope and wire rope slings shall follow code specifications as identified by manufacturer, providing that a safety factor of not less than five (5) is maintained.
E. The eyes of the wire rope slings shall be so formed or spliced as to maintain the safe working load of the sling.
F. Protruding ends shall be covered or blunted.
G. Wire rope shall not be secured by knots, except on haulback lines on scrapers.
H. When "U" bolts are used, they shall comply with code specifications.
I. An eye splice made in any wire rope shall have not less than three full tucks.
J. Except for eye splices, a wire rope shall consist of one continuous piece without a knot or splice.
K. Eyes shall not be formed by using wire rope clips or knots.
L. Wire ropes and wire slings shall be inspected before each use and frequently during use; lubrication is also necessary.
M. Chains and chain slings shall be given a visual inspection before each use and thoroughly inspected every 3 months.
N. Records of inspection are required.
O. Chain links shall be replaced when length increase exceeds 5 percent.
P. After repair, a chain shall be proof-tested.

- Q. A chain shall not be shortened by bolting, wiring, or knotting.
- R. Do not lift hoist if the chain is kinked or knotted.
- S. Maximum load is determined by manufacturers' specifications—use hoisting chain.
- T. The manufacturers' recommendations shall be followed in determining the safe working loads of the various sizes of hooks and rings.

IV. MATERIAL STORAGE
- A. Storage of material shall not create a hazard.
- B. Bags, containers, bundles, etc., stored in tiers shall be stacked, blocked, interlocked, and limited in tier height so that they are stable and secure against sliding or collapse.
- C. Bags shall not be piled more than ten bags high, except when special enclosures are provided. The first five tiers shall be cross-piled and set back with the sixth.
- D. Bulk materials shall not be dumped against a wall higher than stability of the wall will allow. Personnel working in hoppers with high piles of loose material shall be equipped with lifelines.
- E. All commodities shall be stored, handled, and piled with due regard to their fire characteristics and with at least 3 feet between the sprinkler deflector and the top of the stored material.
- F. Commodities shall not be stored within 3 feet of a fire door opening.
- G. Outdoor storage sites shall be free from accumulations of unnecessary combustible materials.
- H. Proper drainage is required.
- I. Clearance signs shall be posted.
- J. The door-locking device on vaults and file rooms shall be of a type enabling a person accidentally locked inside to open the door from the inside.
- K. Storage and handling information is also found under Standard Procedure Instruction No. 12—Hazardous Materials (page 109).
- L. Stored commodities shall not be located under heating equipment using liquid fuel.
- M. "No Smoking" signs shall be conspicuously posted in prohibited areas.

REFERENCES

USDL Title 29 CFR 1504.61, "Gear and Equipment."
USDL Title 29 CFR 1504.62, "Fiber Rope and Fiber Rope Slings."
USDL Title 29 CFR 1504.63, "Wire Rope and Wire Rope Slings."
ANSI Standard A10.2, "Safety Code for Building Construction."
USDL Title 29 CFR 1504.65, "Hooks Other Than Handhooks."
USDL Title 29 CFR 1504.74, "Cranes and Derricks."
ANSI Standard B30.3.0, "Overhead and Gantry Cranes."
ANSI Standard B30.5, "Crawler, Locomotive and Truck Cranes."
ANSI Standard B56.1, "Safety Code for Powered Industrial Trucks."

NFPA no. 505, vol. 10, "Powered Industrial Trucks."
ANSI Standard B20.1, "Safety Code for Conveyors, Cableways, and Related Equipment."
ANSI Standard B30.6, "Safety Code for Derricks."
USDL Title 41 CFR 50-204.3, "Material Handling and Storage."
USDL Title 41 CFR 50-204.75, "Transportation Safety."
NFPA no. 101, "Life Safety Code."
NFPA no. 231, "General Indoor Storage."

STANDARD PROCEDURE INSTRUCTION: NO. 16
Subject: Equipment Guarding for Points of Operation and Machine Drives   Date June 1, 1974

I. MACHINE GUARDING (GENERAL)
   A. Guards will be affixed to the machine where possible and secured elsewhere if for any reason attachment to the machine is not possible. The safeguards will be designed to give maximum operator protection without interfering with the normal operation of equipment.
   B. The point of operation of machines whose operation exposes an employee to injury shall be guarded. The guarding device shall be in conformity with appropriate standards and shall be so designed and constructed as to prevent the operator from having any part of his body in the danger zone during the operating cycle.
   Appropriate specially designed handtools for placing and removing material shall be such as to permit easy handling of material without forcing the operator to place a hand in the danger zone.
   C. When the periphery of the blades of a fan is less than 7 feet above the floor or working level, the blades shall be guarded. The guard shall have openings no larger than ½ inch.
   D. Machines designed for a fixed location shall be properly anchored to prevent them from walking or moving.
   E. Mechanical motions (e.g., rotary, reciprocating) shall be guarded.
   F. Hazardous mechanisms shall be guarded from the actions of cutting and shearing, in-running nip points, screw and worm, and bending and forming.
   G. Hazards involved in machinery operations can be eliminated by application of effective guarding techniques. Examples are: fixed (enclosure) guards, interlocking guards, automatic guards, two-hand tripping devices, ejection devices, feeding tools, and foot control guards.
   H. Every guard shall be reliable in construction, application, and adjustment; e.g., it shall prevent entry of any part of operator's body into the danger zone during operation; it shall allow inspection; it shall present no hazard in itself; it shall contain flying materials.

I. Openings of guards, barriers, or screens at the point of operation shall be small enough to prevent the placing of any parts of the operator's body in the danger zone. Guidelines shall be followed for maximum permissible openings.
   J. Guarding must also take place at the source of power.
   K. All moving parts located so thay any part is 7 feet or less above floor and platform shall be guarded. Included are such items as flywheels; horizontal, vertical, and inclined shafting; pulleys, sprockets, and chains; all belts, collars, couplings, cranks, and connecting rods; and tail rods, extension piston rods, gears, oil openings, friction drives, and projections.

II. SAWS
   A. A circular, hand-fed rip and crosscut table saw shall be guarded by a hood which shall completely enclose that portion of the saw above and in contact with the material being cut. The saw shall also be furnished with a spreader and nonkickback fingers or dogs.
   B. Circular metal saws shall be provided with a guard of not less than $\frac{1}{8}$-inch sheet metal to stop flying sparks.
   C. Sliding cut-off saws shall be provided with an upper hood which shall completely enclose the upper portion of the blade down to a point including the end of the saw arbor. The hood is to be so constructed that it protects the operator from flying splinters, broken saw teeth, etc., and will deflect sawdust away from the operator. The lower exposed portion of the blade shall be guarded by a device which will automatically adjust for the thickness of stock being cut.
   The device shall also be provided with nonkickback fingers, or dogs, and adjustable stops, and it shall return to starting position when released. Rotating blades shall be conspicuously marked on the hood or guard.
   D. All portions of the saw blade on bandsaws shall be enclosed or guarded, except for the working portion between the bottom of the guide rolls and the table. The outside periphery of the guard is to be of solid material.
   Each bandsaw shall be equipped with a tension-control device and shall be provided with a braking device.

III. GRINDING AND SANDING
   A. Guards on abrasive wheels shall cover the spindle end and nut and flange projections. The work rests shall be kept adjusted closely to the wheel, with a maximum opening of $\frac{1}{8}$ inch to prevent the work from being caught between the wheel and the rest.
   Where the nature of the work is such that it covers the side of the wheel, the side covers of the guard may be omitted.

B. Feed rolls of self-feed sanding machines shall be protected with a semicylindrical guard, while drum sanders shall have an exhaust hood or other guard so placed as to enclose the revolving drum, except for that portion of the drum above the table. This procedure also applies to disk sanders, while belt sanders shall be provided with guards at each nip point where the belt runs onto a pulley.

IV. ROTARY SHAPERS, CUTTERS, AND SHREDDERS

A. Planing, molding, sticking, and matching machines shall have all cutting heads and saws (if used) covered by a metal guard which is fastened to the frame carrying the rolls.

B. Hand-fed jointers with horizontal cutting heads shall have an automatic guard which will cover all sections of the head on the working side of the fence. The guard is to adjust itself to cover the unused portion of the head and to remain in contact with the material at all times.

A guard shall be provided which covers the section of the cutting head back of the fence or gauge.

Each jointer with a vertical head shall be provided with an exhaust hood or other suitable guard so arranged as to completely enclose the revolving head, except for a slot for the material to be jointed.

C. Wood shapers and similar pieces of equipment shall have cutting heads enclosed in a cage or adjustable guard. Each spindle shall be provided with a stopping device.

D. Each tenoning machine shall have all cutting heads and saws (if used) covered by metal guards. The operating treadle shall be guarded.

E. Chipper feed systems shall be arranged so that the operator does not stand in a direct line with the chipper spout. The spout shall be enclosed to a height of at least 42 inches from the floor or platform; mirrors shall be installed if the operator is unable to see the chute. Operators shall wear a safety belt attached to a safety-belt line.

F. Shredder, cutter, and duster rotating heads shall be completely enclosed.

G. Rotary cutter guards shall be placed so that a worker cannot reach for material at a point close to the knife; quick power-disconnect devices shall be provided.

H. Turning machines such as lathes shall have all cutting heads covered by a metal guard (hood or shield) that is hinged if adjustments are to be made.

I. Drilling, reaming, boring, and similar machinery shall be provided with safety-bit checks with no projecting set screws.

V. IN-RUNNING NIP POINTS; WORM AND SCREW MECHANISMS
   A. Each mill shall be provided in front and back with safety trip controls.
   B. Each power wringer shall be equipped with a safety bar or other approved guard across the entire front of the feed or the first pressure rolls.
   C. A feeder belt or other effective device shall be provided for starting paper through a calendar stock.
   D. Winders and rewinders (such as those used for paper), with nip points of all drums on the operator's side, shall be guarded with an automatic or manual barrier guard.
   E. Embossers shall be guarded with an automatic or manual barrier guard.
   F. Garnet licker-ins shall be enclosed by covers.
   G. Molders shall be fully guarded with a quick-stop device.
VI. ROTATING AND REVOLVING MECHANISMS
   A. These mechanisms shall be guarded and many require interlocking of the guard with the drive mechanism.
VII. RECIPROCATING MECHANISMS
   A. A scale guard of substantial construction shall be provided at the back of every drop hammer; it shall be so arranged as to stop all flying objects.
   B. Steam hammers shall be provided with a quick-closing emergency valve in the admission pipe line at a convenient location.
   C. A hand protection device is required when the operator is feeding cold material.
   D. On a forging press, a means shall be provided to prevent the ram from moving in either direction when repairs are being made or dies are being changed; oil swabs or scale brushes shall also be provided. A positive type lockout disconnecting switch or valve shall be provided on every mechanical or hydraulic forging press.
   E. Every power press shall be equipped and operated with a point-of-operation guard or a point-of-operation protection device for every press operation performed, except where the point of operation is limited to an opening of ¼ inch or less.
   F. Every power press guard shall be reliable in construction, application, and adjustment. Guards include the following types: die-enclosure, fixed barrier, interlocked barrier, gate or movable barrier, two-head device, pullout device, sweep devices, and pedal cover.
VIII. SHEARING AND CUTTING
   A. Each guillotine-type cutter shall be equipped with controls

which require the operator to use both hands to engage the clutch.
B. The cover over the knife slicers shall be provided with an interlocking arrangement so that the machine cannot operate unless the cover is in place.
C. Cutting knives shall be provided with a hinged hood to cover the knives.
D. Power shears shall be equipped with a positive type lockout device for disconnecting the power.
E. A guard shall be mounted on the stitching head to prevent operators from getting their fingers caught between the stitching head and clincher block.
F. Each sewing machine shall be equipped with an approved guard permanently attached to the machine so that the operator's fingers cannot pass under the needle.
G. When the possibility of flying material exists, a guard shall be provided.

REFERENCES
USDL Title 41 CFR 50-204.5, "Machine Guarding."
ANSI Standard B11.1, "Safety Code for Power Presses."
ANSI Standard B24.1, "Safety Code for Forging and Hot Metal Stamping."
ANSI Standard B15.1, "Safety Code for Mechanical Power-Transmission Apparatus."
ANSI Standard 01.1, "Safety Code for Woodworking Machinery."
ANSI Standard B7.1, "Safety Code for the Use, Care and Protection of Abrasive Wheels."
ANSI Standard P1.1, "Safety in Pulp and Paper Mills."
ANSI Standard L.1, "Textile Safety Code."
ANSI Standard Z50.1, "Safety Code for Bakery Equipment."
ANSI Standard B28.1, "Safety Specifications for Mills and Calenders in the Rubber and Plastics Industries."
ANSI Standard Z8.1, "Safety Code for Foundry Machinery and Operations."

STANDARD PROCEDURE INSTRUCTION: NO. 17
Subject: Hand and Portable Power Tools   Date June 1, 1974

I. GENERAL REQUIREMENTS
A. All hand and portable power tools and equipment furnished by the employee or employer shall be maintained in a safe condition, free of worn or defective parts.
B. All equipment designed to have guards shall have such guards in operation.
C. All portable power driven saws shall be equipped with guards above and below the base plate shoe. The guard shall automatically and instantly return to a covering position.
D. Portable sander safety guards shall have a maximum exposure angle of 180 degrees and shall be located between the operator and the wheel when in use.
E. Automatic shutoff valves are required on all pneumatically powered portable tools. Tool supports are also required.

    F. All electrically powered portable tools having exposed non-current-carrying metal parts of cord and plug connected equipment which are liable to become energized shall be grounded.

II. CONTROL DEVICES

    A. Hand held, power driven woodworking tools shall be provided with a deadman control.

    B. Portable hand held electric tools shall be equipped with switches of a type which must be manually held in closed position.

    C. Sandblasting hoses shall have a deadman control or an effective signal device at the nozzle end.

    D. Compressed air shall not be used for cleaning purposes except where the pressure is reduced to less than 30 pounds per square inch, and then only with effective chip guarding and personal protective equipment.

III. EXPLOSIVE-ACTUATED FASTENING TOOLS

    A. The muzzle ends of all explosive-actuated fastening tools shall have a protective shield or guard at least 3½ inches in diameter, mounted perpendicular to and concentric with the barrel and designed to confine any flying fragments or particles that might otherwise create a hazard at the time of firing.

    B. The mechanism shall be so designed that it cannot be fired during loading, preparation to fire, or if dropped while loaded.

    C. Every lifeline and safety belt shall be of sufficient strength to support a weight of at least 2,500 pounds.

    D. When using a jack, the operator shall make sure the device will support the load. The load rating must be conspicuously marked on the jack.

    E. Hand tools which are unsafe shall not be used.

    F. Wrenches shall not be used when jaws are sprung to the point of slippage.

    G. Mushroomed tools shall not be used.

    H. Wooden handles shall be free of splinters and cracks, and shall be properly placed on tools.

    I. The supervisor shall be responsible for the safe condition of all tools, even those furnished by the employee.

REFERENCES

USDL Title 41 CFR 50.204.4, "Tools and Equipment."
USDL Title 41 CFR 50.204.5, "Machine Guarding."
ANSI Standard O1.1, "Safety Code for Woodworking Machinery."
ANSI Standard B7.1, "Safety Code for the Use, Care and Protection of Abrasive Wheels."
ANSI Standard C1, "National Electrical Code."
ANSI Standard B19, "Safety Code for Compressed Air Machinery."
NFPA no. 70, vol. 5, "National Electrical Code."
USDL Title 29 CFR 1501.23, "Mechanical Point Removers."
USDL Title 29 CFR 1502.72, "Portable Electric Tools."
USDL Title 41 CFR 50.204.8, "Use of Compressed Air."

USDL Title 29 CFR 1501.73, "Hand Tools."
ANSI Standard A10.3, "Safety Requirements for Explosive-Actuated Tools."
ANSI Standard A10.2, "Safety Code for Building Construction."
ANSI Standard B30.1, "Safety Code for Jacks."

STANDARD PROCEDURE INSTRUCTION: NO. 18
Subject: Electrical Wiring, Apparatus, and Equipment   Date June 1, 1974

I. GENERAL CONFORMANCE
   A. Equipment utilizing electricity shall be installed and maintained in conformity with safety codes.
   B. Clearly illustrated instructions for resuscitation of persons suffering from electric shock shall be posted conspicuously in all electrical stations.
   C. Circuits shall be de-energized before any work on them is attempted, and they shall be grounded. Lockout is suggested and tagged.
   D. Standard signs shall be used to indicate the hazard.
   E. Fuse tongs shall be provided and used for changing fuses.
   F. Live parts of electrical equipment operating at 50 volts or more shall be guarded by approved enclosures against accidental contact.
   G. Properly located enclosures shall be used to protect equipment against overcurrent.
   H. All interior wiring systems shall be grounded.
   I. Electrical equipment shall be firmly mounted to various types of surfaces.
   J. Ducts and air-handling spaces shall not be used to transport wiring systems.

II. OUTLET BOXES
   A. Electrical boxes located in damp or wet areas shall be waterproof.
   B. The number of service cords serviced from electrical boxes shall not exceed the number of outlets in the box.
   C. All electrical boxes shall be properly covered and supported.
   D. All electrical boxes shall be wired with a three-wire grounded system.

III. ELECTRICAL GROUNDING
   A. All portable equipment shall use a grounded three-wire extension cord.
   B. All fixed equipment that can be energized shall have an independent, adequately grounded, electrical box.
   C. All electric welding equipment shall be properly grounded, including metal objects within range of the welding power source.
   D. All electrical boxes near damp or wet areas shall have cover plates.

IV. EXTENSIONS
   A. All electric extension cords shall be in proper condition suitable to the situation.
   B. No extension or temporary wiring shall be permitted for any power source carrying more than 3 amperes of current.
   C. All portable electric cords shall be of the three-wire type.
   D. All extension cords shall be equipped with cover (nonconductive) plates over the end of the plug.
   E. All extension cords shall be of the heavy duty type necessary to carry the electrical load of the device.
V. PORTABLE APPLIANCES
   A. Each appliance shall be provided with a means of being disconnected from all ungrounded conductors.
   B. Flexible cord shall be used only in continuous lengths without splice or tap.
   C. Worn or frayed electric cables shall not be used.
VI. GENERATORS, MOTORS, AND TRANSFORMERS
   A. Generators shall be located in dry places; generators carrying more than 150 volts to ground shall not be exposed to accidental contact.
   B. Suitable guards or enclosures shall be provided to protect exposed current-carrying parts of motors unless the motor is designed for the existing conditions.
   C. Exposed non-current-carrying metal parts of transformers shall be grounded.
   D. Oil transformers located indoors shall be installed in a vault of fire resistive construction that is ventilated and well lighted.
   E. Dry type transformers located indoors shall be adequately ventilated.
   F. Unauthorized persons shall be prevented from entering transformer vaults by both physical barriers and posted visible signs.
   G. Materials shall not be stored in transformer vaults.
VII. HAZARDOUS LOCATIONS
   A. Equipment for a specific hazardous location shall not be installed or intermixed with equipment approved for another specific hazardous location.
   B. The three classes of locations are as follows:
      Class I locations—They are those in which flammable gases or vapors are or may be present in quantities sufficient to produce explosive or ignitible mixtures.
      Class II locations—They are those which are hazardous because of the presence of combustible dust.
      Class III locations—They are those which are hazardous because of the presence of easily ignitible fibers or flyings, but in

which such fibers or flyings are not likely to be in suspension in the air in quantities sufficient to produce ignitible mixtures.
C. Special attention must be given for each class of location with reference to meters, instruments, relays, wiring methods, switches, circuit breakers, motor controllers, fuses, motors, generators, lighting fixtures, flexible cords, utilization equipment, signals, and alarms.

VIII. BATTERY ROOMS
A. Storage batteries shall be located so that they are not accessible to other than qualified persons.
B. Batteries of the nonsealed type shall be located in separate rooms or enclosures.
C. Provision shall be made for sufficient diffusion of the gases from the battery to prevent the accumulation of an explosive mixture.
D. Wiring, heating, and ventilation shall meet National Electrical Code specifications.
E. Racks and trays shall be substantial and shall be treated to be resistant to the electrolyte.
F. Floors shall be of acid-resistant construction or be adequately protected from acid accumulation.

REFERENCES
ANSI Standard C1, "National Electric Code."
ANSI Standard C2, "National Electrical Safety Code."
ANSI Standard M28.1, "Safety Procedures for Quarries."
NFPA no. 70, vol. 5, "National Electrical Code."
USDL Title 29 CFR 1501.72, "Portable Electric Tools."

STANDARD PROCEDURE INSTRUCTION: NO. 19
Subject: Operation of Motor Vehicles   Date June 1, 1974

I. COMPANY OWNED OR LEASED MOTOR VEHICLES

Our transportation division shall maintain a transportation supervisor to administer coordinated control over all company owned or leased vehicles. The control office shall have on file a record with a full description of each company owned or leased vehicle. There shall also be kept in this office a driver record card for each individual qualified and approved to operate a company owned or leased vehicle, who is responsible for compliance with the requirements set forth therein and as stated in this Procedure.

II. RULES AND REGULATIONS
A. Company credit cards will be used only for our equipment as necessary. All fueling and servicing is to be done at company shops unless extreme circumstances make the use of credit cards necessary. Any credit card ticket must have the signature of the person purchasing fuel and complete equipment number. If any

of these items are left off of tickets, the charge will be made to the profit center receiving fuel. If a credit card is refused and you have enough personal money with you, and if you are willing to do so, you can pay the bill, get an itemized invoice properly receipted, and submit this invoice to the unit manager for processing to reimburse you. Otherwise, resort to the emergency phone procedure described below.

B. If you have a *major breakdown* or *serious accident* while driving in areas outside the city, you should
   1. On normal work days, call the responsible person assigned by the unit.
   2. After normal business hours and on weekends or holidays, refer to the emergency number provided by the unit manager.

C. As a licensed and an authorized driver of a company owned or leased vehicle, you are covered by the company insurance carrier policy.

D. Our transportation department will maintain an up-to-date register of all vehicle accidents for three years following the date of the accident for examination by the Department of Transportation. The register information will include
   1. Claim number and vehicle number.
   2. Date and hour of accident.
   3. Location of accident, including street address, city, and state.
   4. Number of deaths.
   5. Nonfatal injuries.
   6. Amount of damage in dollars.
   7. Name of driver.
   8. Nature of accident (collision, overturn, fire, cargo damage).
   9. Local or intercity operation.

E. Company owned or leased vehicles are to be used for company business only. Use for personal convenience is not permitted.

F. You are personally responsible for observing all traffic and speed regulations.

G. In case of an accident on the road, observe all state laws which apply. *Do not at any time* drive away from any accident until a police officer has investigated and made his report.

H. Report *all* accidents promptly to the appropriate supervisor of your unit. Report all accidental damage to customer's property during deliveries regardless of amount of damage.

I. Never abandon a company owned or leased vehicle on a road or highway until you have first complied with the emergency procedures indicated and have notified your unit manager or supervisor.

J. Park this vehicle only in authorized places and always lock it when not in use.

K. Company owned or leased vehicles may be driven only by properly authorized company employees. Driving by others is prohibited.
L. Do not overload a vehicle. Stay within the rated tonnage and passenger capacity.
M. All personnel who are assigned vehicles for regular or limited use are responsible for proper care and maintenance of the vehicle while they are using it.

III. DRIVER QUALIFICATIONS
A. All operators must have a currently valid operator's license.
B. An operator must read and speak English, but he does not have to be able to write it.
C. The operator is required to be familiar with methods and procedures for securing cargo in or on the vehicle he operates. However, he is not required to personally perform corrective cargo adjustments or to enter a sealed cargo compartment. The driver should be able to determine by visual inspection and by the vehicle's handling characteristics whether loading is proper so that he can request corrective action if necessary.
D. An operator is required, at least once every twelve months, to prepare and furnish his supervisor with a list of all traffic violations (other than parking) of which he has been convicted or on account of which he has forfeited bond or collateral. If the operator has not been convicted of any violations, he must so certify to his supervisor.
E. All operators of ¾ ton or larger vehicles must have submitted an employment application and successfully completed a road test and written examination and be issued a certificate of such successful completion. The road test must be of sufficient duration to evaluate driving skill and must include testing on the points specified in the Department of Transportation regulations. The form must be signed by the person who administered the test, even if the applicant fails.
F. An operator cannot be permitted to operate a vehicle for three years if he has been convicted of, or forfeited bond or collateral upon, a charge of committing any of the following criminal offenses:
  1. A felony committed while in the driver used vehicle, including his own or another's car.
  2. A crime involving the manufacturing, knowing transportation, knowing possession, sale, or habitual use of amphetamines and narcotics.
  3. Operating a vehicle under the influence of alcohol, amphetamines, or narcotics.

4. Leaving the scene of an accident which resulted in personal injury or death.

The disqualification starts from the date of conviction.

STANDARD PROCEDURE INSTRUCTION: NO. 20
Subject: Handling and Storage of Explosives   Date June 1, 1974

I. GENERAL

The methods and procedures outlined in this instruction are intended to minimize the probability of loss. The specific instructions set forth in each of the identified activity areas concerning explosives are minimal requirements.

II. PERSONNEL

A. Each property facility involved in the storage and/or handling of any materials identified as explosive, regardless of classification, shall maintain at least one supervisory person in attendance who is certified to use and/or supervise the use of explosives.

B. All employees involved in the handling and use of explosives must be a minimum of twenty-one years of age. Prior to any assignment, employees shall be additionally screened for special exposures to determine respiratory ailments, allergies, stability profile, and past performance.

C. Every employee involved in the use and handling of explosives shall be instructed in the safe practices related to arrangement of storage in magazines, handling of containers, transportation from the magazine to the blasting site, preblast surveys, loading of drilled holes, preblast guarding, surplus explosives, misfires, and disposal of empty containers. The instructions shall be in written form posted for employee review. These same instructions shall be reviewed by the loss control coordinator for performance evaluation during facility surveys.

III. CLASSIFICATION OF EXPLOSIVES

A. Class A explosives are those which detonate easily or which have maximum hazard characteristics. Examples are: dynamite, nitroglycerin, black powder, blasting caps, detonating primers, picric acid, lead azide, and fulminate of mercury.

Class B explosives are those which are hazardous because they are flammable. Examples are: solid propellants, cartridge starters, and special fireworks.

Class C explosives include certain types of manufactured articles and devices which contain Class A and/or Class B explosives as components, but in restricted quantities.

- B. Forbidden or not acceptable explosives are those forbidden or not acceptable for transportation by common carriers via rail freight, rail express, highway, or water in accordance with the regulations of the U.S. Department of Transportation.
- C. Those chemicals and fuels identified as having explosive characteristics, although not specifically classified by the U.S. Department of Transportation, shall be stored, handled, and transported in accordance with the most stringent regulatory control and available authoritative information.

IV. STORAGE OF EXPLOSIVES
- A. All explosives, regardless of class, shall be kept in magazines which meet the requirements of Section 495, Chapter 3, of the National Fire Codes, "Storage of Explosives."
- B. Storage magazines for explosives shall be located in accordance with the "American Table of Distances for Storage of Explosives." Class II magazines may be permitted, by authority having jurisdiction, in warehouses and wholesale and retail establishments if the floor of the storage site has an entrance at grade level and the magazine is not more than 10 feet from the entrance.
- C. At construction and blasting sites, Class II magazines for the storage of explosives shall be located at a distance from adjacent buildings, railways, highways, and other magazines. A distance of 150 feet is required between Class II magazines and the work in progress, when the quantity of explosives exceeds 25 pounds. A distance of 50 feet is required for storage of up to 25 pounds of explosives. The authority having jurisdiction shall be consulted for storage requirements at each job site.
- D. Blasting caps, detonating primers, and primed cartridges shall not be stored in the same magazine with other explosives.
- E. Storage magazines for explosives shall be located so they have optimum surface drainage in all directions. The area adjacent to the magazine shall be clear of all combustible materials within a minimum radius of 50 feet. Flora and vegetation with drying-out tendencies will be kept cleared away from the vicinity of the magazine for a minimum distance of a 25-foot radius.
- F. The presence and/or use of open lights, flames, spark-producing devices, firearms, or smoking is prohibited inside or within a 50-foot radius of any explosives magazine.
- G. Any explosives recovered from a misfire must be stored in a separate magazine until disposal criteria are obtained from the manufacturer and/or until the authority having jurisdiction determines the disposal area. Caps recovered from misfires shall not be stored for reuse to initiate any blast.

V. HANDLING AND USE OF EXPLOSIVES
- A. Persons handling explosives shall not smoke or have on their

person or use ignition devices, spark-producing equipment, open flames, or lights. No person shall handle explosives while under the influence of intoxicating liquors, narcotics, or other dangerous drugs.
B. The containers of explosives shall not be opened in any magazine. The original containers or Class II magazines shall be used for taking explosives from the storage magazine to the blasting site.
C. Blasting operations shall be accomplished only during daylight hours except by special permission of the authority having jurisdiction.
D. A twenty-four-hour advance notice to appropriate utility companies shall be given by the person in authority if the blasting is in the vicinity of any utility. The exact location, time schedule, and utilities involved shall be indicated to the appropriate utility representative. A written confirmation is required.
E. The blasting site shall be protected from unauthorized entry. Warning signs, barricades, and signals shall be used. Also, personnel shall be used to direct motor vehicle and/or pedestrian traffic from entering a blast zone where conditions warrant such use. Blasting mats of appropriate size and structural quality shall be used to protect any adjacent exposure.
F. A preblast survey shall be made where the blasting site is in the vicinity of other buildings, dams, utilities, railways, piers, wells, shafts, or tunnels. Special note shall be made of any apparent unusual conditions, and a bench-mark program shall be maintained for the duration of blasting activity in that vicinity. Owners and/or their authorized representatives shall be notified to witness conditions prior to an initial blasting and after blasting operations are completed. The entire bench-mark program shall be a matter of written record.
G. Empty boxes and paper and fiber packing materials which have previously contained high explosives shall not be used again for any purpose, but shall be destroyed by burning at an approved isolated location outdoors. No person shall be permitted nearer than 100 feet after the burning has started.
H. Explosives or blasting equipment that are obviously deteriorated or damaged shall not be used. No explosives, regardless of condition, shall be abandoned. When any explosive has deteriorated to an extent that it is in an unstable or dangerous condition, or if nitroglycerin leaks from any explosive, then the person in possession of such an explosive shall immediately report the fact to the authority having jurisdiction and upon his authorization shall proceed to destroy such an explosive in accordance with the instructions of the manufacturer.
I. All drill holes shall be sufficiently large to freely admit the insertion of the cartridges of explosives. No holes shall be loaded

during drilling operations. Only those holes to be fired in the next round of blasting will be loaded. After loading, all remaining explosives shall be immediately returned to an authorized magazine.
J. The drilling of new holes shall not proceed until all remaining butts of old holes are examined with a wooden stick for unexploded charges, and if any are found, they shall be refired before work proceeds. No person shall be allowed to deepen drill holes which have contained explosives.
K. After loading for a blast is completed and before initiation of the blast, all excess blasting caps or electric blasting caps and other explosives shall immediately be returned to their separate storage magazines.
L. Only electric blasting caps shall be used for blasting operations in congested districts, or on highways, or adjacent to highways open to traffic, except where sources of extraneous electricity make such use dangerous.
M. When a fuse is used, the blasting cap shall be securely attached to the safety fuse with a standard ring-type cap crimper. All primers shall be assembled at least 50 feet from any magazine. Primers shall be made up only as required for each round of blasting.
N. Blasters, when testing circuits to charged holes, shall use only blasting galvanometers designed for this purpose.
O. Only the person making the leading wire connections in electrical firing shall fire the shot. All connections should be made from a bore hole back to the source of the firing current. The leading wires shall remain shorted and will not be connected to the blasting machine or other source of current until the blast is to be initiated.
P. In event of any misfire while using cap and fuse, all persons shall remain away from the charge for at least one hour. If electric blasting caps are used and a misfire occurs, this waiting period may be reduced to thirty minutes. Misfires shall be handled under the direction of the person in charge of the blasting. All wires shall be carefully traced and a search made for unexploded charges.
Q. Explosives shall not be extracted from a hold that has once been charged or has misfired unless it is impossible to detonate the unexploded charge by insertion of a fresh additional primer.
R. Before initiating a blast, a loud warning signal shall be given by the person in charge who has determined that all surplus explosives are in a safe place, all persons and vehicles are at a safe distance or under sufficient cover, and adjacent property is protected.

VI. TRANSPORTATION OF EXPLOSIVES
    A. The transportation of explosives over all highways shall be in

accordance with U.S. Department of Transportation regulations.
B. No person shall smoke, carry matches or any other flame-producing device, or carry any firearms or loaded cartridges in a careless or reckless manner while in or near a motor vehicle.
C. Explosives shall not be carried or transported in or upon a public conveyance or vehicle carrying passengers for hire.
D. Blasting caps, or electric blasting caps, shall not be transported over the highways in the same vehicles with other explosives, except by permit from the authority having jurisdiction.
E. The vehicle used for transporting explosives shall be strong enough to carry the load without difficulty and shall be in good mechanical condition. The vehicle shall have a closed body, tight floors, and any exposed spark-producing metal on the inside of the body shall be covered with wood or other nonsparking materials to prevent contact with packages of explosives.
F. Every vehicle used for transporting explosives shall be marked according to DOT regulations for explosive cargo.
G. The vehicle used for transporting explosives shall be equipped with a minimum of two extinguishers, each having a rating of at least 10BC. Extinguishers shall be inspected daily, determined to be ready for immediate use, and located near the driver's seat.
H. The vehicle used for transporting explosives shall be given the following daily preoperation inspection to determine that it is in proper condition for safe transportation of explosives:
1. Fire extinguishers shall be filled and in working order.
2. All electrical wiring shall be completely protected and securely fastened to prevent short-circuiting.
3. Chassis, motor, pan, and underside of body shall be reasonably clean and free of excess oil and grease.
4. Fuel tank and feed line shall be secure and have no leaks.
5. Brakes, lights, horn, windshield wipers, and steering apparatus shall function properly.
6. Tires shall be checked for proper inflation and defects.
7. The vehicle shall be in proper condition in every other respect and acceptable for handling explosives.

REFERENCES

Institute of Makers of Explosives
420 Lexington Avenue
New York, New York 10017

Bureau of Mines
U.S. Department of the Interior
Washington, D.C. 20242

Bureau of Explosives
2 Penn Plaza
Seventh Avenue and West Thirty-third Street
New York, New York 10001

American Insurance Association (NBFU)
85 John Street
New York, New York 10038

U.S. Department of Transportation
Washington, D.C. 20423

Munitions Carriers Conference, Inc.
1424 Sixteenth Street, NW
Washington, D.C. 20036

STANDARD PROCEDURE INSTRUCTION: NO. 21
Subject: Claims Administration　　　　　　　　　Date June 1, 1974

A. The intent of this procedure is to provide a uniform and uncomplicated method to report losses to the risk management office for prompt attention by the claims administrator. In this connection, one basic universal loss report will be used to report all losses except those involving employee occupational injuries and illnesses. In those instances, only a copy of the employer's first report of injury and/or illness, as required by the state industrial commission, will be furnished to the risk management office.

B. All losses occurring on company property will be reported by the appropriate department head to the risk management department. The only exception will be that of the catastrophe-type loss that occurs on or away from company property after normal hours of operation. Losses of this type will be reported to the security department and relayed to the department head and the risk management department.

C. The following losses shall be reported:
1. MOTOR VEHICLE: Accidents involving company assigned or non-owned vehicles and resulting in physical damage to the automobile, bodily injury, theft, fire, and vandalism.
2. REAL PROPERTY: Damage to company owned, leased, or rented buildings and structures as a result of fire, lightning, windstorm, flooding, or vandalism; damage resulting from the activity of others, such as contractors or vendors.
3. PORTABLE PROPERTY: Losses resulting from theft, fire, or vandalism which involve portable property, such as audiovisual equipment, office machines, art works, and valuable papers.
4. GENERAL LIABILITY: Damage to property or bodily injury to others (visitors, vendors, contractors, inspectors, etc.) resulting from activities or operations of the company.
5. MONEY AND SECURITIES: Loss of money, securities, negotiable certificates, and special event tickets as a result of burglary, armed robbery, or mysterious disappearance.
6. BOILERS AND MACHINERY: Losses involving all steam and hot water boilers and all machinery, including refrigeration equipment with a capactiy of 100 horsepower and over, as a result of explosion, sudden rending, tearing, or damage to these items and their prime movers. Also, damage to all nonrotating electrical switchgear, transformers, power panels, and starters of primary and secondary status, with the exception of those attached to appliances and equipment having a capacity of less than 100 horsepower.

D. All losses are to be reported to:
Risk Management Department
1801 South Jenny Lane
Suite D-8
Temple, Arizona 85281

E. Losses of catastrophic nature involving any fatality and any major structural damage from fire, earthquake, flood, or explosion shall be reported by telephone to:
Risk Management Department
(602) 938-9075
Temple, Arizona
F. This location has 24-hour attendants to receive loss reports and provide claims administration. All telephoned emergency loss reports by employees who are away from company property should be made *collect* to the risk management department. A written loss report shall follow any telephone report to the risk management office.
G. The risk management office will acknowledge receipt of the loss report in written form indicating the loss register number and the basic description of the occurrence. A sample universal loss report form and a sample acknowledgment form are attached for your information.
H. The loss control coordinator shall immediately notify the risk management office of any notice of claim or suit brought against the company relating to an occurrence or injury. He will forward all demands, notices, summonses, or other processes by the claimant or his representative to the risk management office.
I. The risk management department does have draft authority for certain claims such as first party losses, including physical damage losses to autos (collision and comprehensive damage), that do not involve bodily injury or property of others. In the event of a large property loss or damage to third party property and/or bodily injury, we will assign the case to an adjuster and work toward a speedy conclusion. The risk management department will supervise the handling of losses as outlined in this procedure. Risk management personnel should be consulted for answers to questions pertaining to the handling and settlement of any losses not previously identified.

Attachments (2)

SAN PHELCOTT COPPER COMPANY—
ACKNOWLEDGMENT OF LOSS REPORT

Loss Register No.: Date June 1, 1974
Description:

TO:
Processing is now in progress. Please use the attached forms to report any additional important information or to report any future loss.
RISK MANAGEMENT DEPARTMENT
1801 SOUTH JENNY LANE
TEMPLE, ARIZONA 85281

SAN PHELCOTT COPPER COMPANY UNIVERSAL LOSS REPORT			
Department		Department Location	
Report Status ☐ New ☐ Additional Information		Date	Time
Describe the Occurrence			
Persons Involved			
Witnesses			
Police Agencies			
Reported By Date		Reviewed By Date	
Note: If this is an "Additional Information" report please give the date of the loss.			

SAN PHELCOTT COPPER COMPANY
UNIVERSAL LOSS REPORT

Department		Department Location	
Report Status ☐ New ☐ Additional Information		Date	Time

Describe the Occurrence

Persons Involved

Witnesses

Police Agencies

Reported By
Date

Reviewed By
Date

Note: If this is an "Additional Information" report please give the date of the loss.

safety and health protection on the job

The Williams-Steiger Occupational Safety and Health Act of 1970 provides job safety and health protection for workers through the promotion of safe and healthful working conditions throughout the Nation. Requirements of the Act include the following:

Employers: Each employer shall furnish to each of his employees employment and a place of employment free from recognized hazards that are causing or are likely to cause death or serious harm to his employees; and shall comply with occupational safety and health standards issued under the Act.

Employees: Each employee shall comply with all occupational safety and health standards, rules, regulations and orders issued under the Act that apply to his own actions and conduct on the job.

Proposed Penalty: days, or until it is corrected, whichever is later, to warn employees of dangers that may exist there.

The Act provides for mandatory penalties against employers of up to $1,000 for each serious violation and for optional penalties of up to $1,000 for each nonserious violation. Penalties of up to $1,000 per day may be proposed for failure to correct violations within the proposed time period. Also, any employer who willfully or repeatedly violates the Act may be assessed penalties of up to $10,000 for each such violation.

Criminal penalties are also provided for in the Act. Any willful violation resulting in death of an employee, upon conviction, is punishable by a fine of not more than $10,000 or by imprisonment for not more than six months, or by both. Conviction of an employer after a first conviction doubles these maximum penalties.

Inspection: The Occupational Safety and Health Administration (OSHA) of the Department of Labor has the primary responsibility for administering the Act. OSHA issues occupational safety and health standards, and its Compliance Safety and Health Officers conduct jobsite inspections to ensure compliance with the Act.

The Act requires that a representative of the employer and a representative authorized by the employees be given an opportunity to accompany the OSHA inspector for the purpose of aiding the inspection.

Where there is no authorized employee representative, the OSHA Compliance Officer must consult with a reasonable number of employees concerning safety and health conditions in the workplace.

Complaint: Employees or their representatives have the right to file a complaint with the nearest OSHA office requesting an inspection if they believe unsafe or unhealthful conditions exist in their workplace. OSHA will withhold names of employees complaining on request.

The Act provides that employees may not be discharged or discriminated against in any way for filing safety and health complaints or otherwise exercising their rights under the Act.

An employee who believes he has been discriminated against may file a complaint with the nearest OSHA office within 30 days of the alleged discrimination.

Citation: If upon inspection OSHA believes an employer has violated the Act, a citation alleging such violations will be issued to the employer. Each citation will specify a time period within which the alleged violation must be corrected.

The OSHA citation must be prominently displayed at or near the place of alleged violation for three

Voluntary Activity: While providing penalties for violations, the Act also encourages efforts by labor and management, before an OSHA inspection, to reduce injuries and illnesses arising out of employment.

The Department of Labor encourages employers and employees to reduce workplace hazards voluntarily and to develop and improve safety and health programs in all workplaces and industries.

Such cooperative action would initially focus on the identification and elimination of hazards that could cause death, injury, or illness to employees and supervisors. There are many public and private organizations that can provide information and assistance in this effort, if requested.

More Information: Additional information and copies of the Act, specific OSHA safety and health standards, and other applicable regulations may be obtained from the nearest OSHA Regional Office in the following locations:

Atlanta, Georgia
Boston, Massachusetts
Chicago, Illinois
Dallas, Texas
Denver, Colorado
Kansas City, Missouri
New York, New York
Philadelphia, Pennsylvania
San Francisco, California
Seattle, Washington

Telephone numbers for these offices, and additional Area Office locations, are listed in the telephone directory under the United States Department of Labor in the United States Government listing.

Washington, D.C.
1973
OSHA 2003

Peter J. Brennan
Secretary of Labor

U. S. Department of Labor
Occupational Safety and Health Administration

143

Where Does It All End?

Management has been motivated, the supervisors are educated, aware and functioning, the risk manager has his product, and now let's sit back and reap the rewards.—Right?—No, wrong! The really hard work is ahead, but we have the base, the road map or whatever you want to call it. We know where we want to go and have a way to get there. Well—maybe, but more likely, maybe not. The problem with all this machinery we've created to plan, organize, motivate, control, educate, etc., etc., is that it's based entirely on our experience of the past and our knowledge of today. Will all this work tomorrow? Never in a million years—unless it is tempered and constantly changed with the new and the innovative methods that will emerge from your *own* efforts, not to mention the efforts of others.

There is no end to it. Each new action, each new method will suggest and open up new horizons of thought and effectiveness. Neither Loss Control nor Risk Management is a science. Each is an orderly system to accomplish certain objectives. Both are firmly attached to human behavior. What we have tried to do here is to gather the essential elements of loss control systems into an orderly scheme.

We sincerely hope your company benefits by it.

Glossary of Terms

ACCIDENT—An unintended interruption in the orderly processes to attain an objective, sometimes, but not always, involving personal injury and/or property damage. It may result from an unsafe work practice, substandard condition, or combination of both.

ACCIDENT TYPE—A description and classification of the occurrences directly related to the source of injury and an explanation of how that source produced the injury. It answers the question "How did the injured person come in contact with the object, substance, or exposure named as the source of injury, or during what employee activity did the bodily injury occur?"

AMERICAN CONFERENCE OF GOVERNMENTAL INDUSTRIAL HYGIENISTS—ACGIH.

AMERICAN INDUSTRIAL HYGIENE ASSOCIATION—AIHA.

AMERICAN NATIONAL STANDARDS INSTITUTE—ANSI—Originally American Standards Association (ASA), name then changed to United States American Standards Institute (USASI), not ANSI.

AMERICAN SOCIETY FOR TESTING AND MATERIALS—ASTM.

ANHYDROUS—Containing no water.

APPROVED (Method, equipment, etc.)—A method, equipment, procedure, practice, tool, etc., which is sanctioned, consented to, confirmed, or accepted as good or satisfactory for a particular purpose or use by a person or organization authorized to make such a judgment.

AUTOMATIC SPRINKLER SYSTEM—A combination of water discharge devices (sprinklers), distribution piping to supply water to the discharge devices, one or more sources of water under pressure, water flow controlling devices (valves), and actuating devices (temperature, rate of rise, smoke, or other type of device). The system automatically delivers and discharges water in the fire area. There are four basic types of systems:

WET PIPE—All piping is filled with water under pressure which is released by the fusible mechanism in the sprinkler head.

DRY PIPE—All piping contains air under pressure. When a sprinkler head opens, the air pressure is released and the dry pipe valve opens and permits water to flow into the system and to any open sprinkler heads. This system is generally used in areas where piping may be exposed to freezing conditions.

DELUGE—All sprinkler heads are open, and the water is held back at a main (deluge) valve. When this valve is opened by the actuating device, water is delivered and discharged from all heads simultaneously. Generally used where it is desired to wet down a large area quickly, such as an airplane hangar.

PREACTION—Similar to a dry-pipe system; however, air pressure may or may not be utilized. The main water control valve is opened by an actuating device, which permits water to flow to the individual sprinkler heads, and the system then functions as a wet-pipe system. It is generally used in areas where pipe systems are subject to mechanical damage and where it is important to prevent accidental discharge of water.

CARBON MONOXIDE—A colorless, odorless, toxic gas generated by combustion of the common fuels in the presence of an insufficient air supply or where combustion is incomplete. Poisoning is due to the combination of carbon monoxide with the available hemoglobin. This action tends to exclude oxygen and so suffocates the victim by lowering the ability of the blood to carry oxygen. First aid: reduce the loss of heat from the body and supply oxygen as rapidly as possible by moving the victim to fresh air; administer artificial respiration if breathing has stopped, or preferably, administer oxygen or oxygen mixed with 5 to 7 percent carbon dioxide.

CATASTROPHE—Loss of extraordinarily large dimensions in terms of injury, death, damage, and destruction.

CERTIFIED SAFETY PROFESSIONAL (CSP)—An individual who has been certified by the Board of Certified Safety Professionals as having achieved professional competence by remaining abreast of the technical, administrative, and regulatory developments in his chosen field and who maintains professional integrity in his relations with clients, associates, and the public that reflects the highest standard of ethics.

CONDUCTIVITY—The property of a circuit which permits the flow of electricity. It is the reciprocal of resistance and is measured by the ratio of the current flowing through a conductor to the difference of potential between its ends.

CONSENSUS STANDARD—A standard developed according to a consensus agreement, or general opinion, among representatives of various interested or affected organizations and individuals.

CONSTRUCTION (TYPES OF BUILDING)—There are five common types of building construction classified according to certain fire-safe characteristics: fire-resistive, heavy timber, noncombustible, ordinary, and wood frame.

FIRE-RESISTIVE CONSTRUCTION—A broad range of structural systems capable of withstanding fires of specified intensity and duration without failure. Common fire-resistive components include masonry load-bearing walls, reinforced concrete or protected steel columns, and poured or precast concrete floors and roofs. Although fire-resistive structures do not, in themselves, contribute fuel to fire, combustible trim, ceilings, and other interior furnishings may produce an intense fire and pose a serious threat to life safety.

HEAVY TIMBER CONSTRUCTION—Construction characterized by masonry walls, heavy timber columns and beams, and heavy plank floors. Although not immune to fire, the great mass of the wooden member slows the rate of combustion.

NONCOMBUSTIBLE CONSTRUCTION—Includes all types of structures in which the structure itself (exclusive of trim, interior finish, and contents) is noncombustible but not fire resistant. Exposed steel beams and columns and masonry, metal, or asbestos panel walls are the most common forms.

DEPARTMENT HEAD—The group directly responsible to the middle management group for final execution of policies and implementing directives by rank-and-file employees, and for attainment of objectives in assigned organizational units through practices and procedures approved and issued by top or middle management.

DEPARTMENT OF HEALTH, EDUCATION AND WELFARE—HEW.

DEPARTMENT OF LABOR—DOL.

DEPARTMENT OF TRANSPORTATION—DOT.

DISABLING INJURY—American Standard Z16.1: "An injury which prevents a person from performing a regularly established job for one full day (24 hours) beyond the day of the accident." National Health Survey: "A bed disabling injury is one which confines a person to bed for more than half of the daylight hours on the day of the accident or on some day following. A restricted activity injury is one which causes a person to cut down on his usual activities for a whole day." General definition: Bodily harm which results in death, permanent disability, or any degree of temporary total disability. OSHA-BLS Standard defines these as "lost work-day cases." This is an occupational injury or illness as a result of which (1) the employee would have worked but could not, or (2) the employee was assigned to a temporary job, or (3) the employee worked at a permanent job less than full time, or (4) the employee worked at a permanently assigned job but could not perform all duties normally assigned to it.

DRY-CHEMICAL EXTINGUISHER—An extinguisher containing a chemical which extinguishes fire by interrupting the chain reaction wherein the chemicals used prevent the union of free radical particles in the combustion process so that combustion does not continue when the flame front is completely covered with the agent. Three types of base chemical agents are used: sodium bicarbonate, potassium bicarbonate, and ammonium phosphate (multipurpose). These are used primarily on Class B and C fires; however, multipurpose dry chemicals are also effective on Class A fires.

EAR DEFENDERS—Plugs or muffs designed to keep noxious noise from the ear to preserve hearing acuity.

EMPLOYEE—General term for an employed wage earner or salaried worker. Used interchangeably with "worker" in the context of a work situation, but a worker is not an employee when he is no longer on the payroll. Any person engaged in activities for an employer from which that person receives direct payment for services performed.

EMPLOYER—General term for any individual, corporation, or other operating group who hires workers (employees). The terms "employer" and "management" are often used interchangeably when there is no intent to draw a distinction between owners and managers.

ENFORCEMENT—The exercising of executive power or the use of authority, direct or delegated, to require the adherence to prescribed standards, policies, rules, and regulations.

EXPERIENCE RATING (MERIT RATING)—Process of basing insurance premiums on the employer's own record—as in workmen's compensation, unemployment insurance, and commercially insured health and insurance programs—so that he may benefit from a good record.

EXPLOSIMETER—A device which detects and measures the presence of gas or vapor in an explosive atmosphere.

EXPLOSIVE:

LIMITS—The minimum (lower) and maximum (upper) concentration of vapor or gas in air or oxygen below or above which explosion or propagation of flames does not occur in the presence of a source of ignition. The explosive or flammable limits are usually expressed in terms of percentage by volume of vapor or gas in air.

RANGE—The difference between the lower and upper flammable (explosive) limits, expressed in terms of percentage by volume of vapor or gas in air.

EXPLOSIVE—Any chemical compound or mechanical mixture that is used or intended for the purpose of producing an explosion. Contains oxidizing and combustive units or other ingredients in such proportions, quantities, or packing that an ignition by fire, by friction, by concussion, by percussion, or by detonation of any part of the compound or mixture may cause such a sudden generation of highly heated gases that the resultant gaseous pressures are capable of producing destructive effects on contiguous objects or of destroying life or limb.

FIRE PROTECTION ENGINEERING—The field of engineering concerned with the safeguarding of life and property against loss from fire, explosion, and related hazards. It is concerned with integrated programs involving the design and use of structures, equipment, processes, and systems, including the areas of prevention, detection and alarm, and fire control and extinguishing, and gives consideration to functional, economic, and operational factors.

FLAMMABLE—Any substance that is easily ignited, burns intensely, or has a rapid rate of flame spread. (Although this word is usually an adjective, in loss control contents it is used as a noun.) Flammable and inflammable are identical in meaning; however, the prefix "in" indicates negative in many words and can cause confusion. Flammable therefore is the preferred term.

FLASH POINT—The lowest temperature of a liquid at which it gives off sufficient vapor to form an ignitible mixture with the air near the surface of the liquid or within the vessel used. The flash point can be determined by the open cup or the closed cup method. The flash point determined by the open cup method is usually somewhat higher than that obtained with the closed cup method. The latter is commonly used to determine the classification of liquids which flash in the ordinary temperature range.

FUSIBLE LINK—A link which usually consists of two pieces of metal joined by a solder with a low melting point. It is manufactured at various temperature ratings and subject to varying normal maximum tension. When it is installed and the ratio temperature is reached, the solder melts, and the two metal parts separate; this initiates desired actions, e.g., signaling the presence of fire, automatically closing doors, actuating extinguishing equipment, or releasing the water from a sprinkler head.

GAS—A state of matter in which the material has a very low density and viscosity, can expand and contract greatly in response to changes in temperature and pressure, easily diffuses into other gases, and readily and uniformly distributes itself throughout any container. A gas can be changed to its liquid or solid state only by the combined effect of increased pressure and decreased temperature (below the critical temperature).

GOGGLES—Spectacles or glasses to protect the eyes against dust, impacting objects, strong light, or other harmful environmental influences.

GUARDRAIL—A device consisting of post and rail members, or of wall section; it is erected to mark points of major hazards and to prevent individuals from coming in contact with these hazards.

HANDRAIL—A rail designed to assist personnel in using stairways or as a guard around floor and wall openings. A flight of stairs having four or more risers should have a standard handrail as specified in ANSI Standard A12, "Safety Code for Floor and Wall Openings, Railings, and Toeboards." Any guardrails that are installed should also conform to this standard.

HAZARDOUS MATERIALS—Materials possessing a relatively high potential for harmful effects upon persons.

ILLUMINATION—The amount of light flux a surface receives per unit area. It may be expressed in lumens per square foot or in footcandles. The rate at which a source emits light energy, evaluated in terms of its visual effect, is spoken of as light flux and is expressed in lumens.

JOB HAZARD ANALYSIS—The breaking down of any method or procedure into its component parts to determine the hazards connected therewith and the requirements or qualifications of those who are to perform it. A method for studying a job in order to (1) identify hazards or potential accidents associated with each step or task and (2) develop solutions that will eliminate, nullify, or prevent such hazards or accidents.

LIQUEFIED PETROLEUM GAS (LP GAS)—A compressed or liquefied gas usually comprised of propane, some butane, and lesser quantities of other light hydrocarbons and impurities; obtained as a by-product in petroleum refining. It is used chiefly as a fuel and in chemical synthesis.

LOSS RATIO (INSURANCE)—A fraction calculated by dividing the amount of losses by the amount of premiums and expressed as a percentage of the premiums. Various bases are used in calculating the loss ratio, e.g., earned premium loss ratio, written premium loss ratio, etc.

MIXTURE—A combination of two or more substances which may be separated by mechanical means. The components may not be uniformly dispersed.

MOTOR VEHICLE ACCIDENT—Any accident involving a motor vehicle in motion that results in death, injury, or property damage. However, motion of the motor vehicle is not required in a collision between a railroad train or a streetcar or another motor vehicle.

NATIONAL FIRE PROTECTION ASSOCIATION (NFPA)—An organization of individuals or groups interested in the prevention of damage by fire.

NATURAL GAS—A combustible gas, composed largely of methane and other hydrocarbons with variable amounts of nitrogen and noncombustible gases, obtained from natural earth fissures or from driven wells. It is used as a fuel, in the manufacture of carbon black, and in chemical synthesis of many products. It is a major source of hydrogen for the manufacture of ammonia.

NEGLIGENCE—The failure to do what reasonable and prudent persons would do under similar circumstances, or failure to do what reasonable and prudent persons would have done under the existing circumstances.

OCCUPATIONAL ILLNESS AND INJURY—Illness is any abnormal physical condition or disorder, other than one resulting from an occupational injury, caused by exposure to environmental factors associated with employment. It includes acute and chronic illness or disease which may be caused by inhalation, absorption, ingestion, or direct contact. An occupational injury is any injury such as a cut, fracture, sprain, amputation, etc., which arises in or out of the course of performing assigned work.

OCCUPATIONAL INJURY AND ILLNESS RECORDS—OSHA—Records of each reportable occupational injury (including fatality) and illness which are required by every employer covered by the National System for Uniform Recording and Reporting of Occupational Injury and Illness.

OCCUPATIONAL INJURY OR ILLNESS, REPORTABLE—OSHA—Any disability or permanent impairment to an employee which results from any exposure in the work environment that either (1) results in death, or (2) prevents the employee from performing his normal assignment during the next regular or subsequent work day or shift, or (3) not causing death or loss of time, *(a)* results in transfer to another job or termination of employment, or *(b)* requires medical treatment other than first aid, or *(c)* results in loss of consciousness, or *(d)* is diagnosed as an occupational illness, or *(e)* results in restriction of work or motion.

OCCUPATIONAL SAFETY AND HEALTH ADMINISTRATION—OSHA.

OCCURRENCE—An incident often classified as relatively major or minor. In insurance, occurrence is distinguished from accident by the fact that it is apparent or foreseen as occurring as the result of certain activities.

ORGANIC—Designation of any chemical compound containing carbon. (Some of the simple compounds of carbon, such as carbon dioxide, are frequently classified as inorganic compounds.) To date nearly one million organic compounds have been synthesized or isolated. Many occur in nature; others are produced by chemical synthesis. In medicine, producing or involving alteration in the structure of an organ; opposed to functional.

POTENTIAL—That which can, but has not yet, come into being; latent; unrealized.

PREEMPLOYMENT EXAMINATION—A physical examination of a job applicant conducted prior to his employment.

PRODUCT LIABILITY—The liability a merchant or a manufacturer may incur as the result of some defect in the product he has sold or manufactured, or the liability a contractor might incur after he has completed a job from improperly performed work. The latter part of product liability is called "completed operations."

RADIATION—The emission and propagation of energy in the form of waves through space or through a material medium. Usually refers to electromagnetic radiation such as gamma rays and ultraviolet rays, and may also apply to alpha and beta particles or heat waves.

RELIEF VALVE—A valve designed to release excess pressure within a system without damaging the system itself.

RESPIRATOR—A protective device for the human respiratory system designed to protect the wearer from inhalation of harmful atmospheres. There are two types of respiratory protective devices: (1) air purifiers, which remove the contaminants from the air by filtering or chemical absorption before inhalation, and (2) air suppliers, which provide clean air from an outside source or oxygen from a tank.

SAFETY BELT—A life belt worn by telephone linemen, window washers, etc., attached to a secure object (telephone pole, window sill, etc.) to prevent falling. A seat or torso belt securing a passenger in an automobile or airplane to provide body protection during a collision, sudden stop, air turbulence, etc.

SAFETY HATS—Rigid headgear of varying materials designed to protect the workman's head not only from impact, but from flying particles and electric shock, or any combination of the three. Safety helmets should meet the requirements of American Standard Z89, "Standard for Industrial Protective Helmets."

SERIOUS INJURY—(Z16.1) The classification of a work injury which includes (1) all disabling work injuries and (2) nondisabling injuries in the following categories: *(a)* eye injuries from work-produced objects, corrosive materials, radiation, burns, etc., requiring treatment by a physician, *(b)* fractures, *(c)* any work injury that requires hospitalization for observation, *(d)* loss of consciousness (work related), and *(e)* any other work injury (such as abrasion, physical or chemical burn, contusion, laceration, or puncture wound) which requires treatment by a medical doctor, or restriction of work or motion or assignment to another regularly established job.

STANDARD PROCEDURE INSTRUCTION—A series of logical steps by which all repetitive business action is initiated, performed, controlled, and finalized. A procedure establishes what action is required, who is required to act, and when the action is to take place. A procedure is also a medium for communicating managerial policy decisions and for determining the routine by which a specific operation will be performed.

STATIC ELECTRICITY—Electric charges that are normally stationary. Nonflowing electricity commonly generated by friction.

STODDARD SOLVENT—A dry-cleaning solvent possessing a flash point of 100°F (40.6°C), which evaporates without residue and consists of aliphatic saturated materials and, in some formulations, 15 to 20 percent of aromatics. The fire hazard is about that of kerosene, but it is a more satisfactory cleaning solvent. It is available under a number of trade names.

SUBSTANDARD CONDITION—Any physical state which deviates from that which is acceptable, normal, or correct in terms of its past production or potential future production of personal injury and/or damage to property or things; any physical state which results in a reduction in the degree of safety normally present. It should be noted that accidents are invariably preceded by unsafe acts and/or unsafe conditions. Thus, unsafe acts and/or conditions are essential to the existence or occurrence of an accident.

SUBSTANDARD WORK PRACTICE—A departure from an accepted, normal, or correct procedure or practice which has in the past actually produced injury or property damage or has the potential for producing such loss in the future; an unnecessary exposure to a hazard; conduct reducing the degree of safety normally present. Not every unsafe act produces an injury or loss, but, by definition, all unsafe acts have the potential for producing future accident injuries or losses. An unsafe act may be an act of commission (doing something which is unsafe) or an act of omission (failing to do something that should have been done).

SUPERVISOR—Any individual held responsible for the behavior and production of a group of workers.

SURVEY—Inspection and comprehensive study or examination of an organization, environment, or activity for insurance and/or accident prevention purposes.

THRESHOLD LIMIT VALUES (TLV)—Values of airborne concentrations of substances; represents conditions to which it is believed that nearly all workers may be repeatedly exposed, day after day, without adverse effect. At these concentrations, the worker may continually breathe in a substance for eight hours per day without harm. Because of wide variation in individual susceptibility, exposure of an occasional individual at or even below the threshold limit may not prevent discomfort, aggravation of a preexisting condition, or occupational illness. Threshold limits should be used as guides in the control of health hazards and should not be regarded as fine lines between safe and dangerous concentrations. The American Conference of Governmental Industrial Hygienists (ACGIH) adopts a list of threshold limit values (TLV) each year for more than 450 substances.

TOEBOARDS—A guard commonly installed around flywheels and other equipment in open pits and on overhead catwalks. The installation of toeboards should conform to ANSI Standard A12.1, "Safety Requirements for Floor and Wall Openings, Railings, and Toe Boards." Toeboards should be at least 4 inches high and should be made of wood, metal, or metal grille not exceeding 1-inch mesh. Toeboards at flywheels should be placed as close to the edge of the pit as possible. Wooden toeboards for permanent installations should be of 1-inch by 4-inch stock or heavier.

TOXIC SUBSTANCE—OSHA—A substance that demonstrates the potential to induce cancer, to produce short and long term disease or bodily injury, to affect health adversely, to produce acute discomfort, or to endanger life of man or animal resulting from exposure via the respiratory tract, skin, eye, mouth, or other routes in quantities which are reasonable for experimental animals or which have been reported to have produced toxic effects in man.

APPROXIMATE LETHAL CONCENTRATION—LCca—A concentration of a chemical substance which has been reported either to produce a lethal effect or to be the maximum concentration that could be tolerated without death during an exposure of one day.

APPROXIMATE LETHAL DOSE—LDca—A dose of a chemical substance which has been reported either to produce a lethal effect or to be the maximum dose which could be tolerated without death.

LETHAL CONCENTRATION FIFTY—LC50—A calculated concentration which, when administered by the respiratory route, would be expected to kill 50 percent of a population of experimental animals during an exposure of four hours.

LETHAL DOSE FIFTY—LD50—A calculated dose of a chemical substance which is expected to kill 50 percent of an entire population of experimental animals exposed through a route other than respiratory.

NONTOXIC—Failing by experience and/or experiment to result in physiologic, morphologic, or functional changes adversely affecting the health of man or animal according to reasonable and appropriate tests.

VAPOR

DENSITY—Weight of a vapor per unit volume at any given temperature and pressure.

PRESSURE—The pressure exerted at any given temperature by a vapor either by itself or in a mixture of gases. It is measured at the surface of an evaporating liquid.

VARIANCE—The quality, state, or fact of varying; the degree of change or difference. In mathematics, the degree of "scatter," variability, or divergence about the mean of a distribution, computed as the square of the standard deviation. It also applies to a license granted by governments to pollute beyond acceptable limit in return for promises and plans on curbing pollution.

Index

Index

A

Accident reporting:
  cargo losses, 74
  injuries: first-aid, 74
    forms, 64–67
    serious or disabling, 74
  motor vehicle accidents, 74, 131
    form, 69
  property damage, 74
ACGIH (American Conference of
    Governmental Industrial
    Hygienists), 105
Administration:
  assignment of, 14

Administration (*Cont.*):
  claims (*see* Claims administration)
  corporate risk management, 17
  loss control director, 13, 15-20
    duties of, 18-20
    qualifications of, 20
  loss control objectives, 14-15
Air contaminants, 105-106
American Insurance Association, 137
Anhydrous ammonia, 113
ANSI (American National Standards Institute), 76-77, 99, 102, 105, 109, 113, 116-118, 121, 122, 126-128, 130
Apparel and appliances, safety, 76-77
  eye protection, 76, 82
  head protection, 76
  hearing protection, 77
  protective clothing, 76-77
Atmospheres, dangerous, testing of, 106
Automotive maintenance area, standards for, 77-78

B

Bureau of Explosives, 137
Bureau of Mines, 137

C

Cargo losses, reporting, 74
Chemical laboratories, 79-87
  cork and stopper borers, 85
  eating and smoking in work areas, 82

Chemical laboratories (*Cont.*):
  electrical equipment, 83–84
  eye protection, 82
  flame hazard, 81–82
  handling glassware, 84–85
  hazardous operations, 80
  hood operations, 85–86
  housekeeping, 79–80
  laboratory manipulations, 86–87
    (*See also* Chemicals, handling of)
  machinery and powered equipment
    hazards, 80
  personal protective equipment, 82–83
  reporting procedures, 81
  respiratory protection, 83
  responsibility for, 79
  safety rules, 87
  vacuum equipment, 83
  working after regular hours, 82
Chemicals, handling of, 87–93
  compressed gas cylinders, 88–89,
    109–110, 117
  disposal of waste materials, 87–88
  drying of solvents, 90–91
  flammable liquids, 89–90, 110, 111
  pyrophoric materials, 91–93
Claims administration, 138–140
  form for loss report, 140
Clothing, protective, 76–77
Color coding, 108–109
Combustible liquids and materials,
  110–112, 117
Compliance program, 52–70
  investigation of losses, 54
  loss summary analysis, 54–55
  occupational safety and health
    standards, 53–54
  performance evaluation (*see*
    Performance evaluation)

Compliance program (*Cont.*):
  rules and regulations, 52
  standard procedure instructions, 53, 71–140
  traffic laws, 53
Compressed air, 117–118
Compressed Gas Association, 118
Compressed gases, 88–89, 109–110, 117

D

Dip tanks for flammable or combustible liquids, 112–113
Director (*see* Loss control director)

E

Educational program, 21–39
  assignment and demonstration, 22
  for developing supervisory skills, 23–39
  orientation and instruction, 22–23
  performance review, 23
  preemployment physical examination, 22
  progressive educational instruction, 23
  recruitment of personnel, 22
  (*See also* Training)
Egress, means of, 99–102
    exit doors, 100–101
    exit facilities, 99–100
    exit signs, 100
    exits: general-purpose, 101
      high-hazard, 101
      ramps, 101–102

Electrical wiring, apparatus, and
    equipment, 128–130
  appliances, portable, 129
  battery rooms, 130
  in chemical laboratories, 83–84
  equipment rooms, 78
  extensions, 129
  generators, 129
  grounding, 128
  hazardous locations, 129–130
  motors, 129
  outlet boxes, 128
  transformers, 129
Emergency call list, 73
Emergency reporting, 73
Employees, responsibility of, for
    compliance program, 7–8
Employers, responsibility of, for
    compliance program, 6–7
Engineering program, 40–51
  rearrangement and modification, 41
  research and development, 41
  surveys and analysis (*see* Loss control,
    survey examples)
Environmental control (*see* Occupation-
    al health and environmental
    control)
Equipment guarding (*see* Guarding)
Exits (*see* Egress, means of)
Explosives, 113, 133–137
  classification of, 133–134
  handling of, 134–136
  personnel, 133
  storage of, 134
  transportation of, 136–137
  use of, 134–136
Extinguishers, fire, 114–117
Eye protection, 76, 82

F

Facilities:
  people, 28–29
  processes, 27–28
  specifications, 26–27
Federal Register, 3, 7
Fire protection, 114–117
  alarms, 116
  extinguishers, portable, 114–115
  sprinkler systems, automatic, 115–116
  standpipes and hoses, 116
First-aid injuries, reporting, 74
Flammable liquids and materials, 110–112

G

Generators, 129
Guarding, equipment, 122–126
  cutters, 124
  cutting, 125–126
  grinding, 123–124
  machine, general, 122–123
  nip points, 125
  reciprocating mechanisms, 125
  rotating and revolving mechanisms, 125
  sanding, 123–124
  saws, 123
  shapers, 124
  shearing, 125–126
  shredders, 124
  worm and screw mechanisms, 125

H

Hazardous materials, 109–113
  anhydrous ammonia, 113
  compressed gases, 88–89, 109–110, 117
  explosives (*see* Explosives)
  flammable and combustible liquids, 110–111
    dip tanks for, 112–113
  liquefied petroleum gases, 113
  spray finishing using flammable and combustible materials, 111–112
Head protection, 76
Health (*see* Occupational health and environmental control)
Hearing protection, 77
Hoisting equipment, 103–104
Hoses and standpipes, 116
Housekeeping:
  in chemical laboratories, 79–80
  general, 75–76

I

Injuries, reporting: first-aid, 74
    forms, 64–67
    serious or disabling, 74
Inspection procedures, 8–9
Institute of Makers of Explosives, 137

L

Ladders, 96–98
Liability claims, form, 68
Liquefied petroleum gases, 113
Load-bearing surfaces, 96
Loss control, 9–10
  corporate policy for, 13–14
  definition of, 25–26
  management control for, illustrated, 27
  objectives of, 14–15
  performance standards, 28–29
  survey examples, 43–51
    arrangement of facilities, 44–45
    blood services, 47
    conclusion and recommendations, 49–51
    drugs and narcotics, 47
    fire prevention, 45
    nature of operations, 43–44
    professional exposure, 46–47
    special control areas, 48–49
  (*See also* Loss prevention)
Loss control director, 15–20
  desired qualifications of, 20
  duties of, 18–20
Loss control policy, 13–15
Loss experience, 61–70
  automobile accidents, form, 69
  liability claims, form, 68
  measuring effectiveness, 62–63
  occupational injuries and illnesses, forms, 64–67
  property losses, form, 70

Loss prevention, approach to, 26-29
    facilities, 26
    illustrated, 27
    people, 28
    processes, 26-28

M

Maintenance activities, 77-79
    automotive maintenance area, 77-78
    equipment rooms, electrical and mechanical, 78
    power machinery and equipment, 77
    welding operations, 78-79
Management participation, 11-20
    department head, 12-13
Manhole covers, 95-96
Manlifts, 104-105
Material handling, 118-122
    equipment, 118-121
    operators, 118
Material storage, 121
Mechanical equipment rooms, 78
Motor vehicles:
    accident reporting, 74, 131
    operation of, 130-133
        company-owned or leased, 130, 132
        driver qualifications, 132-133
        rules and regulations, 130-132
Munitions Carriers Conference, Inc., 137

N

National Electrical Code, 77
National Safety Council, 11, 18

NFPA (National Fire Protection Association), 102, 105, 109, 113, 114, 116, 117, 122
Noise exposure, 107-108

O

Occupational health and environmental control, 105-109
  air contaminants, 105-106
  color coding, 108-109
  noise exposure, 107-108
  radiation hazards, 108, 109
  testing of dangerous atmospheres, 106
  ventilation, 106-107
Occupational injuries and illnesses, forms, 64-67
  (*See also* Accident reporting)
Occupational Safety and Health Act of 1970, 4-10
  employees, responsibilities of, 7-8
  employers, requirements for, 6-7
  inspection procedures, 8-9
Openings, floor, 95-96

P

Performance evaluation, 14, 53, 55-61
  rating form, 56-57
  rating form guideline, 57-61
Platforms, 102-105
  hoisting equipment, 103-104
  powered, 102
  roof car, 102-103
  working, 103

Policy, loss control, 13-15
Poster, compliance, 5
    illustrated, 143
Power tools, hand and portable, 126-128
    control devices, 127
    explosive-actuated fastening tools, 127
Procedures (*see* Standard Procedure Instructions)
Property damage, reporting, 74
Property losses, form, 70

R

Radiation hazards, 108, 109
Railings:
  specifications for, 94-95
  types of, 94
Risk categories, areas of activities, 9-10
Risk management personnel, 10

S

Safety committee, 49
Scaffolds, 98-99
Sprinkler systems, automatic, 113, 115-116
Standard Procedure Instructions, 53, 71-140
  (*See also* specific practice)
Standpipes and hoses, 116

Supervisor:
  condition report of, form, 31
  corrective measures by, 32
  developing supervisory skills, 23-39
  identifying responsible conditions, 30
  improving job methods, 33-34
    illustrated, 34
  investigating conditions, 29
    illustrated, 30
  measured performance, 32
    illustrated, 33
Surfaces, walking and working, 93-99
  aisles and passageways, 93-94
  ladders, 96-98
  loading, 96
  openings, 93-96
  railings for, 94-95
  scaffolds, 98-99
  toeboards, 95
  wet, 93-94

T

Toeboards, 95
Tools (*see* Power tools)
Traffic laws, 53
Training:
  controlling actions, 37-39
    illustrated, 38, 39
  preparation for, 34-37
  progressive steps, 37
    illustrated, 37
  (*See also* Educational program)
Transformers, 129

U

U.S. Department of Labor, 5-8,
  50, 98, 109, 121, 122
U.S. Department of Transportation, 137

V

Ventilation, 106-107

W

Walking and working surfaces (*see*
  Surfaces)
Welding operations, maintenance
  standards for, 78-79